AF722501

Discovering Abundant Health

A Divine Engineering Approach to Organ Systems and Nutrition

John P. Wiens

Inspired Resources

Published by Inspired Resources, LLC
Phoenix, Arizona

ISBN 979-8-9960794-0-7 (paperback)
ISBN 979-8-9960794-1-4 (eBook)

First Edition, 2026

This work represents years of independent research and writing by the author. AI tools assisted with select imagery and organizational tasks. All content, research, analysis, and conclusions are the author's own.

Cover design by Ron Ruvalcava

To my dear and loving mother, who throughout life was a pillar of faith and protector in prayer. Inspired by her strength and support, I always knew nothing was impossible. This one is for you, Mom.

Table of Contents

List of Figures vii

List of Abbreviations ix

Preface xi

Introduction 1

My Story 2

Part 1 – Basic Guide to Reversing Metabolic Issues 5

Proactive Steps 6

A. Stop consuming 7

B. Eating Right 8

C. Strategic Supplements 10

Trauma and Stress 10

Part 2 – Organs, Nutrition, and Systemic Health 13

Electrolyte Balance and Implications for Systemic Health 14

Iron Balance, Copper, and Implications for Systemic Health 19

Phospholipids and Implications for Systemic Health 23

Brain and Central Nervous System 25

Cardiovascular System 33

Respiratory System 37

Kidneys and Implications from Liver Health 41

Pancreas 44

Spleen and Lymphatic System 46

Thyroid Function and Implications from Liver Health 48
Adrenal function, hormones, and implications for liver health 49
Gastro-intestinal Tract 52
Liver and Gallbladder 57
Senescent Cells and Low-Grade Chronic Inflammation 62
Part 3 – Select Macros, Micros, and Other Nutrients 65
Protein – Not Just for Bodybuilders and the Young 65
Intriguing Roles of "Conditionally Essential" Amino Acids 67
Look, up in the cells, it's omegas.....it's phospholipids! 71
Taurine and Choline – Brothers in Arms 73
The Fantastic Trio – Magnesium, B1, & B3 75
Vitamin C – The Unsung Hero of Brain and Tissue Health 80
The Exotic Benefits of Zinc 81
TUDCA – A Bear of a Supplement 82
The Amazing Benefits of Berberine 83
Part 4 – East Meets West: 85
Traditional Chinese Medicine 88
In the Beginning 88
Lineages, Myths and Legends 90
The Advent of TCM 93
Qi (Vital Energy) 94
Yin and Yang 94
Five Elements 95

Theory of Organs 96
Acupuncture Meridians, Pulse Taking, and Herbs..................101
Ancestral Diet and Dairy (or lack of)...104
The Levant, Patriarchs, and Ancient Israel 105
Historical Introduction..105
Geographical Context 108
Heroic Faith and Epic Sagas...110
The Testaments of the Twelve Patriarchs 112
The Patriarch's Lifestyle and Diet 116
Ancient Israel Diet Overview...118
Biblical Dietary Wisdom..120
Part 5 – Integrative Concepts and Health Modes 129
Summarizing Intro 129
Metabolic Syndrome and Select Nutrient Flow Charts...................131
Cholesterol Metabolism..132
Taurine and Choline Pathways 136
The Great Cheeseburger Conundrum...140
Bread 140
Dairy...142
Meat 143
Holistic Engineering and the Metabolic Organ Pyramid................145
Chicken-or-the-Egg (Liver or Gut) 145
The Metabolic Organ Pyramid..147

More Traditional Wisdom....150
Health Impacts of Exercise Via mTOR and AMPK Pathways....153
Stress and Sleep....157
Stress....157
Sleep....160
Alternative Health Modes....164
Chiropractic....164
Sauna and Cold Plunge....165
Red Light Therapy....167
Grounding or Earthing....168
Part 6 – Nutritional, Organ, and Health Summaries....171
The Most Important Foundational Nutrients....171
Tier 1: The Master Regulators....172
Tier 2: The Essential Regulators....186
Tier 3: The Specialized Essentials....194
Tier 4: The Supporting Cast....204
Health Strategies for Organs, Ailments, and Systemic Health....211
Cardiovascular Disease and Disorders....211
Diabetes & Metabolic Syndrome....214
Cancer – Prevention & Support....217
Neurodegenerative Disease and Nerve Damage....220
Lungs & Respiratory Ailments....222
Bone & Muscle Impairment....226

Immune System: Spleen, Lymphatic System & Thymus 229
Kidney Disease 232
Pancreatic Ailments 235
Thyroid Dysfunction 237
Adrenal-HPA Dysregulation 239
Gut & Microbiome – The Root 242
The Liver – Master Regulator (& Gallbladder) 246
Summary Conclusion 251
Part 7 – Conclusionary Roundup – Tying It All Together 253
My Story – continued 253
Did You Know? 254
One More Intriguing Cycle – The Longevity-Vitality Key 257
The Process – CoQ10 and Mitochondrial Engine of Longevity. 257
Mitophagy: The Longevity Key 259
The Top 12 Foods – A comprehensive Review 261
Deeper Connections 263
Key Take-away Strategies 265
In Closing 267
Afterword 269
References 273

List of Figures

Figure 1. Amino Acids and Metabolites Biosynthesis 70

Figure 2. Major Hormones and Energy Production 71

Figure 3. Five Elements Controlling and Productive Cycles 95

Figure 4. Five Organs Controlling Cycle 98

Figure 5. Productive Cycle Associations Between Emotions and Organs 100

Figure 6. The Patriarchs Diet and Ancient Israel Diet 119

Figure 7. Pathways for Good Cholesterol, Bile, & Healthy Outcomes 132

Figure 8. Nutrient Pathways for Healthy Hormone & Bile Metabolism 134

Figure 9. Nutrient Pathway for Healthy lbLDL Cholesterol Management 134

Figure 10. Nutrient Pathways for Methylation, Cysteine, & Choline. 137

Figure 11. Nutrient Pathway for Choline and Vagus Nerve Regulation 137

Figure 12. Gut-Liver-Brain Axis-Triangle 146

Figure 13. Metabolic Organ Pyramid (MOP) 148

Figure 14. Metabolic Spectrum Guage 151

Figure 15. Exercise Spectrum Guage 155

Figure 16. Sleep Cycle 160

Figure 17. The Mitochondrial Engine 257

List of Abbreviations

AA — Amino Acid

AMPK — AMP-activated Protein Kinase

ANS — Autonomic Nervous System

ATP — Adenosine Triphosphate

BAT — Brown Adipose Tissue

BBB — Blood-Brain Barrier

CKD — Chronic Kidney Disease

CNS — Central Nervous System

COPD — Chronic Obstructive Pulmonary Disease

CRE – Circuit Resistance Exercise

CVD — Cardiovascular Disease

DHA — Docosahexaenoic Acid

DNA — Deoxyribonucleic Acid

EE — Endurance Exercise

FXR — Farnesoid X Receptor

GABA — Gamma-Aminobutyric Acid

HGH — Human Growth Hormone

HIIT — High-Intensity Interval Training

HPA — Hypothalamic-Pituitary-Adrenal Axis

IR — Insulin Resistance

LDL — Low-Density Lipoprotein

LGCI — Low-Grade Chronic Inflammation

LPS — Lipopolysaccharide

MOP — Metabolic Organ Pyramid

mTOR — Mammalian Target of Rapamycin

MTHFR — Methylenetetrahydrofolate Reductase

NAC — N-Acetylcysteine

NAD — Nicotinamide Adenine Dinucleotide

NAFLD — Non-Alcoholic Fatty Liver Disease

NS — Nervous System

PC — Phosphatidylcholine

PE — Phosphatidylethanolamine

PG — Phosphatidylglycerol

PNS — Peripheral Nervous System

PS — Phosphatidylserine

PTSD — Post-Traumatic Stress Disorder

REE — Resistance Exercise

REM — Rapid Eye Movement

SAM — S-Adenosylmethionine

TCM — Traditional Chinese Medicine

TMAO — Trimethylamine N-oxide

TSH — Thyroid-Stimulating Hormone

TUDCA — Tauroursodeoxycholic Acid

VEE – Vigorous Endurance Exercise

WAT — White Adipose Tissue

Preface

I never intended to write this book. For years, I was content solving complex engineering problems and the various other tasks my interdisciplinary career required, but health was something I always assumed was ok — until metabolic dysfunction spiraled into a crisis that conventional approaches couldn't resolve. Doctors offered symptom management and vague solutions. I refused to accept it.

What followed was years of deep research into nutritional science, organ physiology, metabolic pathways, and ancient health wisdom. As an engineer, I approached my health the same way I'd approach any complex problem: methodically and rigorously, insisting on understanding the "why" behind the "what" in order to synthesize the bigger picture of how everything works together for solutions.

But here's what frustrated me: the information existed — scattered across peer-reviewed journals, ancient texts, traditional practices, and modern research — but no one had connected it all. Diet books told me what to eat without explaining biological mechanisms. Medical texts described organs in isolation without showing how systems interconnect. Holistic practitioners referenced ancient wisdom without scientific rigor. Biblical dietary laws were dismissed as arbitrary rather than understood as divine engineering principles.

I needed a comprehensive roadmap that integrated everything. It didn't exist. So I built it.

Discovering Abundant Health integrates over 280 scientific/cited references with biblical wisdom, Traditional Chinese Medicine, and historical insights modern research is only beginning to validate. It applies systems thinking to reveal how organ function, nutritional biochemistry, and divine design intersect in ways most health books never explore. Whether you see God's hand in human biology or not, the scientific evidence and insights will deepen your understanding. I pray this book is a fruitful journey and joy to read.

The book progresses through seven parts — Part 1: practical tips for healing metabolic issues. Part 2: organ systems and liver's central metabolic role. Part 3: select nutrients and their targeted benefits. Part 4: biblical principles and TCM bridged with current research. Part 5: traditional and modern approaches, exercise, and lifestyle. Part 6: comprehensive nutrient catalog for ongoing reference. Part 7: conclusionary wrap-up with concept and action take-aways.

You can read straight through for comprehensive knowledge or as a reference, consulting sections as needed. Health practitioners will find the scientific rigor valuable. Health enthusiasts and faith-based readers will appreciate how ancient wisdom intersects with modern research and how biblical principles align with physiological design. Anyone with chronic health issues will find the clarity they need.

Health isn't a luxury; it's the foundation for pursuing what truly matters. The roadmap is in your hands. Abundant health awaits.

Introduction

It is said, the G.I. tract or gut is like our second brain. Even stranger, the trillions of microbes and bacteria in our gut, both good and bad, actually communicate with our brain, but also affect the gut barrier, revealing health risks for issues ranging from simple indigestion and sleep issues to·depression, metabolic disease, and dementia. We're told, "everything begins with the gut", which is true, but since the G.I. tract cannot function without resilient tissue and microbiome, healthy conjugated bile, and robust enzymes to digest and absorb nutrients, ailments may actually begin with the liver.

A wise person once divinely said "Ask and it will be given; seek and you will find; knock and the door will be opened." — Matthew 7:7. I take this to heart, not only as a spiritual invitation, but as a mental and motivational decree as well. I believe God invites us to take life's lemons and make lemonade – and 3.5 years ago, I got quite a lemon! What we do with these crises govern our path, so with that, I invite you along for an intriguing journey of health insights and discovery.

This book details insights not only from my personal health journey regarding the gut-liver-brain axis, but also from a holistic systems engineering-like approach; consequently, revealing how the liver, apart from the brain, is truly the king of all organs and master regulator of total metabolic health. Along the way, a comprehensive survey of all the organ systems, nutrients, and ancient nutritional wisdom is explored, syncing up with modern cutting-edge science.

I'm neither a doctor nor biochemist, but with a background in engineering, research, and a life-long keen interest in nutrition, I've always been able to dig deep and solve complex problems when inspired. With proper nutrition and a healthy lifestyle, the body is amazingly designed to heal itself. The pages in this book reveal concepts, compelling data, and possibly life-changing information gleaned from over three years and thousands of hours of research.

Initially, my research began with videos of naturopathic doctors and their experience with clinical results. This quickly led to deeper dives into scientific research in order to gain a better understanding of the interactive roles between organs and nutrients and how it all ties together, including intriguing paths into ancient cryptic history.

In seven parts, this book conveys several genres including basic concepts with practical tips; followed by a longer, more academic deep dive into organs and nutrition; and finally, more targeted nutritional, historical, holistic, and integrative sections connecting dots and pulling together all the data into more simple conclusions.

My Story

I've always considered myself fortunate to be blessed with good health. Aside from chronic burping since my early 20s, I feel like my system has always been able to withstand whatever was thrown at it. Although my diet was slightly better than average, by my late 40's to early 50's, increased indigestion, heart palpitations, and sleep issues began to creep in, despite a fairly active life style.

Upon rushing to the ER at age 56 for acute symptoms of anxiety, impending doom, pounding heart, sweating, shortness of breath, brain fog, and gut inflammation to rule out any heart issues, a follow-up exam revealed not only an inflamed duodenum and fatty liver, but also fatty pancreas. Later analysis shows it was even worse.

Six years prior, I went to the ER for a gallbladder attack, but declined surgery because I wanted to try and resolve it naturally. Initially, I improved my diet, but then gradually slipped back into a typical diet until this situation occurred, which to date, has been the biggest challenge I've ever faced in my life, among many.

Despite my dilemma, it was difficult convincing the doctor that I was facing a serious ordeal. Since all the blood tests displayed results within normal ranges (except bilirubin), my ordeal was minimized within the context of overall decent health, so not until I began researching, along with expertise from my acupuncturist, were insightful connections made. In hindsight, I realized that the origins of this crisis had been brewing under the surface for decades.

In Chinese medicine, liver Qi stagnation is at the root of many metabolic diseases. Along with poor diet, genetics, and for reasons explained later, a person can also become predisposed to digestive, liver, gallbladder, and eventually pancreatic and other issues via certain traumatic or chronic emotional stressors, especially early in life. I believe this to be true in my case, but regardless of yours, let us now embark on a voyage of incredible discovery to better health.

Part 1 – Basic Guide to Reversing Metabolic Issues

For thousands of years people have lived out healthy lives largely free of metabolic disease provided they had access to a sufficiently balanced diet and were able to evade hostile environments from war and infectious disease. Today we face much different challenges.

Our bodies are uniquely designed to thrive from the many food sources God has provided for us on this earth, but in modern times, we've lost the common sense and folk knowledge our ancestors had in consuming a healthy diet including the way it was grown, raised, fermented, prepared, etc.

The western processed food diet combined with the many chronic stress factors present in modern society have wreaked havoc on our health, especially in the United States. It's as if, for every ailment or nutrient deficiency caused by highly processed foods, there arises a pharmaceutical solution (or masking of symptoms really) to offset the detriment caused….hmmm.

Today, more than ever, people are looking for answers to health and wellbeing, and while many excellent resources are available, wading through the mountains of data can be overwhelming. While many well-respected experts may be accurate on the vast majority of what they espouse, they might also be well-intentioned, but mistaken on key issues as well, so it's important to evaluate and judge each topic on its own merits according to the science, historical evidence, and triangulation of many sources. I've tried my very best to do just that.

Although not practical to cover every condition or deficiency that exists, in an effort to analyze, synergize, and summarize loads of complex data and bio-mechanisms into practical tips, this book strives to be as thorough and concise as possible; addressing key metabolic junctions for systemic health; while also providing a comprehensive grasp beginning with some basics here in Part 1.

Thus, here's what I've discovered about healing the liver, and by extension, an unhealthy gut, gallbladder, and pancreas which in time, can deliver a host of cascading benefits downstream to the central nervous system and other vital organs through not only dietary changes, but by also addressing suppressed anger/trauma, chronic stress, and negative emotions.

Proactive Steps

Here are some of the basic, but most important general actions that can be taken to begin reclaiming health, listed in order of relevance:

A. First, stop consuming bad stuff – This may sound obvious, but the right supplements can only help if we stop ingesting the "poisons" that cause problems in the first place.

B. Eating the right foods – The optimum is to try and get as many nutrients from whole foods as possible because they have all of the micro-nutrients and other quantum level substances in the right synergistic ratios from nature that are impossible to replicate in even the best supplement.

C. Whole food supplements – If possible, taking supplements derived from food/herb concentrates rather than synthetic nutrients and isolated non-chelated minerals are far better.

An expansion of these steps, backed by research and listed in more detail below, along with references, are actions we can consider for long-term health whether in a health crisis or not. An epidemic 40% of the U.S. adults has fatty liver, 15% has kidney disease, and close to 60% has some sort of digestion issues or insulin resistance, not to mention CVD. Allowing for overlap, that approaches 85%!

Cancers, cardiovascular diseases, diabetes, neurological disorders, and dementia are often the end stage of chronic inflammation that may go undetected for years from asymptomatic, but impaired gut and liver issues. More visible signs such as accelerated aging and degenerative joint/skeletal issues can follow as well, so we should take care to improve our health and lifestyle sooner rather than later if we want to live out healthy, productive lives. The following claims are based on cited references, but also backed by science as well:

A. Stop consuming

1. "Vegetable" oils (seed/corn/soy/etc.) – Found to be a major cause of inflammation and fatty liver/stagnation, especially with frying or combined with sugar or processed foods. Research suggests it's the third leading source of all-cause mortality. Far worse than sugar! I would avoid as much as possible. (1. Axe, 2017) (2. Berg, 2022) (3. Eckberg, 2021)

2. High fructose corn syrup – 20x more likely to cause fatty liver via insulin resistance and inflammation than other sugars! – not to mention weight gain. The liver can process only small amounts every day (±50 grams), so depending on health, small amounts of fruit or moderate amounts with activity, especially berries, but definitely avoid corn syrup as much as possible. (4. Berg, 2019) (5. Ekberg, 2019)

3. White/processed flour and sugars – Leads to IR (insulin resistance) and/or inflammation, which can lead to diabetes, fatty liver, accelerated aging, etc. Avoid especially during acute health conditions. (6. Pam, 2018) (7. Berg, 2019)

4. Grain fed dairy/meats/eggs, and unregulated farm-raised fish contain unhealthy high ratios of omega-6 fats vs omega-3s found in grass-fed/pasture-raised diets, leading to more inflammation throughout the body. (8. Gundry, 2022)

B. Eating Right

1. For gastritis, leaky gut, and other acute gut issues –
   Avoid inflammatory foods listed above, and temporarily, spicy foods, most grains, nuts, and dairy. Include foods like bone broth, cabbage/sauerkraut, cruciferous and root vegetables, garlic, polyphenol foods like olive oil, berries, leafy greens, grass-fed meat and wild-caught fish.
   (9. Berg, 2020) (10. Axe, 2016) (11. Gundry, 2021)

Nutrients necessary for maintaining healthy G.I. tissue include glutamine, glycine, arginine, zinc, vitamin A, and probiotics. (12. Berg, 2020)

2. For fatty liver and gallbladder issues – Foods high in Choline/PC, taurine, cysteine & methionine like eggs, grass-fed dairy/meat, cruciferous vegetables, and fish help to heal the liver. TUDCA and bile salt precursor foods such as artichoke, beets, turmeric, and milk thistle improve gallbladder function. (13. Berg, 2019) (14. Ekberg, 2022) (15. Berg, 2018)

3. Eating the right grains – While a huge debate rages over whether whole grains are beneficial, research shows that ancient grains like Kamut, einkorn, and barley are more nutritious and anti-inflammatory than modern wheat.

   However, even modern wheat is healthy and gluten tolerable when freshly milled as a whole grain and leavened with yeast or sourdough to reduce so-called "anti-nutrients" like lectins and phytates. Almost half of the nutrition is lost in one day and 90% in three days after milling, however, if not baked immediately, or at least frozen to minimize loss. More on this will be covered in Part 5. (16. Brian St. Pierre, Settling the great grain debate., n.d.)

4. Anti-inflammatory foods in general / Good dairy (17. Axe, 2017) (18. Gundry, 2017)

C. Strategic Supplements

Countless great supplements (and bad) exist on the market today, but for brevity, only a few key examples which are either deemed strategically impactful or exhibit great benefits across multiple systems are listed here as it pertains to this study. As always, you are encouraged to do your own research.

1. Magnesium (19. Berg, 2018)
2. Whole-food Multivitamin/mineral (20. QNL, n.d.)
3. Whole-food Vitamin C (21. Livingood, 2022)
4. Whole-food Zinc (22. Axe, 2023)
5. Key "conditionally essential" amino acids
    - Glutamine (23. Osborne, 2019)
    - Taurine (24. DeLauer, 2024)
    - Tyrosine (25. Childs, 2023)
    - Glycine – (26. Berg, 2024)
6. TUDCA (27. Berg, 2021)
7. Berberine (28. Berg, 2023)

Trauma and Stress

Research has shown that aside from diet, another risk factor for liver/gallbladder ailments and other metabolic disease is childhood trauma and suppressed rage, as our internal stress responses are said to be formed by the age of five. People with this type of background can be at a disadvantaged health risk, so taking proactive steps to compensate before mid-life are encouraged.

Adult PTSD experiences by veterans, assault victims, and others can also result in liver and gallbladder damage. Interestingly, Chinese medicine has long referred to anger, stress, and irritability as major sources for liver stagnation and has also linked the stomach to worry and anxiety as well. (29. Spa-Doctor, 2019) (30. Shaw, 2019)

Although advice is given in media regarding emotional stress and therapy, this pathway is rarely addressed as it pertains to the liver, apart from dietary solutions. In this scenario, as will be discussed later, chronically stressed adrenal glands via elevated adrenal hormones appear to preempt the liver as the initial source for this pathology, but the adrenals are also regulated by the liver in a feedback loop, so nutrition is very important in either case.
(31. Jin-Yong Joung, 2019)

In short, the human body can repair itself and attain metabolic equilibrium given the right nutrients and lifestyle, but it doesn't happen overnight. It takes years or decades to reach an unhealthy state, so it may take months or even years to regain a more resilient, healthier state of wellness; and while there are no guarantees in this life, I believe the effort is well worth it, so please be encouraged!

The information here is not meant to replace a doctor's advice and treatment, especially when in a crisis or critical state, but rather is for augmentation and general guidance toward achieving long-term health goals. Please seek medical help if a crisis arises or persists. With that, let's dive into Part 2 describing how all the systems work.

Part 2 – Organs, Nutrition, and Systemic Health

The following discussion and analysis detail a rigorous effort to compile and summarize complex data and relevant research into a more practical and concise format by connecting dots and painting a broad picture of how all major bodily functions and organs are regulated downstream of liver and gut health.

An entire career could be devoted to writing a functional medicine almanac, however, this book was inspired to highlight not only important well-known topics, but to also seek out crucial, but rarely addressed topics as well. Discussion on organ systems is presented from a bottom-up perspective regarding downstream metabolic effects, working backwards up the chain toward the gut and liver.

Part 2 is slightly more academic than the following parts, but should really help to instill confidence in the research presented and lay a foundation of familiarity for the various organs, nutrients, and processes referenced in later sections. However, if it appears too complex or in depth, feel free to skip ahead to other parts and refer back later to topics of interest when more details are needed.

In addition to featuring many of the amazing nutrients our Creator has provided – several of which have become popular buzzwords today – certain intriguing, but critical "underdogs" are also explored. Fitness industry topics will be left largely to the abundance of public resources already available, but will be reviewed as well in Part 5.

In conducting my research, I found it curious how overcaution with certain nutrients by conventional medical wisdom can be resolved simply by healthy regulation rather than restriction in most cases. For instance, dietary needs for nutrients like salt, iron, and calcium, tend to be downplayed or outright discouraged because of adverse effects experienced by many people in poor metabolic health today.

Another factor in this bias is that most people probably do get enough of these nutrients because of how processed foods are often fortified with B vitamins, iron, calcium, excess salt, etc., although not in the bioavailable forms that we need. In fact, if not for fortification of processed foods, the mortality rate would probably surge like it did 100 years ago when white flour first came out.

While many other nutrients are rightfully highlighted in nutritional science, the aforesaid ones are often taken for granted until switching to an unfortified whole foods diet. That said, whole food and food-derived nutrients are truly medicine if consumed properly and consistently. Hydration and sleep are also key pillars for health.

Electrolyte Balance and Implications for Systemic Health

As complex and crucial components to health, fluid and electrolyte balance are often taken for granted and underappreciated, but are essential for life and vibrant health, especially as we get older. Like many regulating processes, electrolyte homeostasis has its origins in the liver and maintains it through the kidneys. Many nutrients, hormones, and enzymes work together to regulate hydration.

Angiotensinogen is a peptide hormone that gets synthesized and released first from the liver, and then along with feedback from the kidney enzyme renin, produces angiotensinogen I to the lungs and angiotensinogen II to the adrenals which then use cholesterol to produce a steroid hormone called aldosterone. Aldosterone regulates blood pressure via fluid balance from potassium and sodium regulation through the kidneys. (32. Scott JH, 2023)

The thyroid is also involved in maintaining electrolyte balance. A parathyroid hormone known as PTH promotes absorption of magnesium through the intestines and regulates release and re-absorption through the bones and kidneys. In turn, PTH production is affected by magnesium levels in a feedback loop.

As one of the most important minerals involved with over 300 known reactions in the body, magnesium is also responsible for converting food to energy, creating proteins, DNA repair & formation, muscle function, heart regulation, and neurotransmitters along with being an essential electrolyte. (33. Cleveland Clinic, 2021)

PTH and magnesium together regulate Vitamin D metabolism, which in turn, along with potassium, regulates calcium homeostasis. Sufficient salt, magnesium, vitamin D, and potassium intake can therefore help to mitigate calcification, as deficiencies can lead to unregulated calcium and chronic electrolyte imbalances which increase the risk of higher blood pressure, kidney failure, strokes, heart attacks, and heart failure over the long run.

A low salt diet is often recommended by the medical industry in these cases, but ironically, since salt deficiency depletes magnesium which PTH relies on, via increased urination, a low salt diet in addition to intestinal malabsorption and/or chronic deficiency of these minerals over time may be the cause. (34. DiNicolantonio, 2021)

Since salt induced thirst is a more recognized response to fluid imbalance and lack of salt is not, it's important to stay hydrated and get enough salt in the diet, especially for athletes, older people with a reduced response mechanism, and people on whole food diets.

In fact, the old paradigm of 1500 - 2300 mg has been shown to be far too low for most people without hypertension and current research suggest at least 3000 and upwards of 4000 or even more depending on activity level and potassium intake. This equates to a minimum of 1 1/3 teaspoons/day or more for people adding salt to a whole foods diet. (35. Medical News Today, 2016)

Although most people do get plenty of salt through the typical western diet, research has shown that a low salt diet for people without hypertension carries the same mortality risks as someone on a high sodium diet *with* hypertension. Ironically, one of the first go-to remedies during many hospital visits are saline IVs. Go figure.

Preserving enough of the other electrolytes like potassium and magnesium is also extremely important because salt deficiency can lead to wasting of these minerals which need to be pulled from the bones and muscles.

High salt danger for unhealthy people with cardio risk may be less a factor of salt levels as it is from low potassium and/or calcification of the arteries which prevent them from flexing in order to regulate blood pressure. Many experts believe the sodium to potassium ratio is more important than the actual reduction of salt, so increased potassium through dietary foods is beneficial since many people are deficient anyway. (36. Berg, 2019)

Blood osmolality is also a major factor that influences the signaling mechanisms that regulate fluid balance. Blood osmolality refers to the concentration of mainly salt, glucose, and urea in the blood. The thirst mechanism, reduced in elderly who don't recognize dehydration from gradually increased osmolality, can be triggered by consuming salt which rapidly increases concentration.

Prior to thirst, osmotic sensors in the brain direct the hypothalamus to secrete arginine vasopressin (AVP) into the pituitary gland, which is then released into the blood stream to suppress water excretion and signal increased reabsorption in the kidneys.

As dehydration increases, these osmotic sensors send a thirst signal via glutamate to the brain to consume more water, whereas GABA inhibits thirst and supports overall signal regulation. GABA, is an important neurotransmitter biosynthesized from glutamate via glutamine that acts as a signaling molecule for several organs and processes in the body. (37. Li S, 2023)

GABA is activated by cholecystokinin (CCK) to quench thirst and is crucial for regulating not only fluid balance signaling, but also positive mediating effects on sleep quality, the pancreas, G.I. tract, pituitary gland, adrenals, and sex organs along with possible treatments for rheumatoid arthritis.

Another pathway assisting with this fluid balance mechanism is the renin-angiotensin system (RAS) which begins in the liver and directs kidneys to reabsorb sodium similarly to the way AVP signals reabsorption of water. (38. Wang J, 2023) (39. Takashi Matsuda, 2020)

As one can see, electrolyte homeostasis is very complex and involves several organs, especially the kidneys. Low salt can lead to dehydration if adequate water is not consumed to offset increased urination caused by a lack of water-retaining capacity from sodium depletion in the kidneys.

In addition, low salt can not only stress adrenals by over-producing aldosterone and cortisol to regulate this capacity, but also lead to mineral wasting and result in the loss of a whole range of health benefits like blood sugar control, better sleep, lower stress hormones, improved skin conditions, and better digestion.

In summary, difficulty in regulating electrolytes could be a sign of not only mineral deficiencies, gut malabsorption, and kidney dysfunction, but also a more long-term condition of liver damage, which affects downstream functionality of the adrenal, thyroid, and kidney's ability to regulate them.

Iron Balance, Copper, and Implications for Systemic Health

Apart from red blood cell production and premenopausal health, the importance of iron is often minimized because of the many fortified processed foods available, and for people with metabolic disfunction and excess iron accumulation in the wrong places.

It's not practical to cover the functions for every single nutrient; however, it was eye opening to learn just how foundational iron regulation is for many of the body's processes including DNA synthesis and repair, metabolic homeostasis, hormones, muscle growth, and bone health. Iron is especially important for people switching to whole food diets, who if not eating the right foods, may suffer subclinical deficiencies without even knowing it.

A major player in the regulation of iron is a peptide called hepcidin. Hepcidin is produced in the liver for central regulation of iron homeostasis in the body. Along with a regulation feedback loop from iron itself, hepcidin production is triggered by inflammation and dysregulation by iron in various body tissues including the heart, brain, kidneys, lungs, etc., as a way to sequester excess iron in the liver, but can also induce anemia if too little from stores are released. (40. Sangkhae V, 2017)

An impaired liver can cause dysregulation of hepcidin and therefore iron levels in the blood that are either too high or too low in other body tissues and organs.

Since excess iron has no pathway for excretion, gut health is crucial for regulating iron uptake in the gut. (41. Subramaniam, 2017) Whereas iron overload has been associated with metabolic disfunction, conversely, some of the consequences resulting from iron deficiency, in addition to anemia, include:

1. Muscle weakness and impaired muscle growth. Intense exercise in general reduces nutrients like B vitamins and minerals such as sodium, potassium, magnesium, calcium, selenium, and zinc (in order); while weight lifting in particular, requires more Vitamin A & D, B vitamins, Vitamin C & E, zinc, iron, magnesium, and calcium for recovery (multiple sources).
2. Restless leg syndrome could be an indication of deficiencies in iron as well as magnesium, B12, and Vitamin D (multiple sources).
3. Sunken eyes/cheeks may signal deficiencies in iron and copper and can also affect vision.
4. Shallow breathing and low oxygen may be caused by low iron levels.
5. Dysregulation of heart and heart failure (42. Szklarz M, 2022)

Copper and Vitamin A also regulate iron in the body, so iron deficiency, if not from lack of consumption or malabsorption in the gut, can result from copper/Vitamin A deficiency or disruption of iron homeostasis in the liver as mentioned above.

Copper is primarily stored in the liver and kidneys and the liver regulates how much is stored or sent to the rest of the body. Blocked bile ducts or gallbladder can prevent excretion of copper leading to copper toxicity from the liver and a deficiency (i.e., iron) and other symptoms in the rest of the body. However, when excess is too great, copper will be excreted through the kidneys through an alternate, or AMPK pathway. (43. Angove, 2023)

AMPK is an essential enzyme for triggering autophagy or cell regeneration in the body and provides an alternate pathway for liver/bile regeneration (apart from TUDCA, the most powerful liver/gallbladder supplement). Exercise and certain foods strongly activate AMPK.

AMPK, which supports a more catabolic pathway as opposed to the anabolic pathway of mTOR associated with new muscle and tissue growth, also promotes energy balance and regulation of copper toxicity in the liver and redirects the path away from various stages that can lead to cancer like initial tumor growth. However, research reveals that the suppression effect of AMPK may be minimized once cancer has been rooted. More on AMPK in coming sections.
(44. Xie, 2022) (45. Jillian Levy, 2018) (46. Diana Vara-Ciruelos, 2019)

Lectins in legumes (beans, peanuts, soybeans), spinach whole grains, and other foods can interfere with the absorption of iron, calcium, phosphorus, and zinc. Even worse, they can damage the intestinal wall, leading to leaky gut and all sorts of resulting issues if foods not baked, fermented, boiled, or pressure cooked properly.

The most important iron absorbing foods are vitamin C/fermented foods, and meat which stimulate iron uptake by inhibition of hepcidin at the intestinal wall. The most important iron inhibitors are phytic acid/phytate foods like grains, resveratrol, polyphenols, tannins, calcium, and foods like milk, and eggs. (multiple sources)

Since healthy dairy like aged cheeses, kefir, and yogurt are both fermented and milk products, the implication is that by supporting both sides, they may be able to help regulate iron uptake. Too many oxalates, which appear in many iron containing foods, may also influence uptake, as they can impair gut tissue.

It's probably not a good idea, for example, too consume too much raw spinach on a regular basis, but rather, cook it instead. Moderate amounts of curcumin may help regulate iron, but too much may decrease iron availability. These are some examples that illustrate the case for eating a balanced diet.

In addition to eating foods listed above, additional support from AMPK provides resiliency to the system and is activated by exercise, a low sugar diet, and high polyphenol containing foods such as berries, dark chocolate, red grapes/wine, tea, nuts, and olives/oil.

These help promote iron homeostasis and other benefits by supporting and helping to heal the liver, which in turn, supports balance in other parts of the body. In summary, iron homeostasis is elemental in regulating many of the body's structural support components and metabolic processes as well as hormonal feedback.

Phospholipids and Implications for Systemic Health

Essential fats have gained more recognition in recent years and while omega-3s have become the star of the show, and rightly so, their quiet hulking relative phospholipids (PLs) have gone largely unnoticed in media. PLs commercially isolated and combined together for food additives are known as lecithin, but are often extracted using chemicals, so natural sourcing is key.

Discovered over a century ago, current research is revealing more health benefits of phospholipids. As explained in following sections, PLs are indispensable building blocks for cell membranes and signaling molecules, not only for the brain, but for the whole body.

Research also shows that adequate PL intake promotes organ health, mitochondria, muscle, and longevity. PL composition between various PL types varies according to organs and tissues and can be either biosynthesized in the body or consumed from food. (47. Yucan Dai, 2021) (48. Heden, 2016)

The most abundant PLs in mitochondria are phosphatidylcholine (PC) and phosphatidylethanolamine (PE) at 40% and 30% respectively, followed by phosphatidylinositol (PI) and cardiolipin (CL) accounting for 10-15% together, phosphatidylserine (PS) and phosphatidic acid (PA) at 5% each, and finally, only about 1% for Phosphatidylglycerol (PG). PL composition by organ however, tells a different story. (49. Osman C, 2011)

PE is most abundant in three of the most vital organs, with its largest concentration in the brain, but is also the main component in the heart and kidneys at 55%, 49%, and 48% respectively; followed by PC as the second largest PL in these organs at 31%, 38%, and 35% respectively. PE is also stated to be a whopping 90% in the G.I. tract as well. (50. Choi J, 2017) (51. Wnorowski, 2021)

Accounting for only 1% in the body, PG comprises its largest concentration in the lungs at 11%, while the liver is the only vital organ where PC dominates, at 48%. The relative amounts of each PL type in various tissues implies the importance for their respective organ functions as will be shown in each organ section of this book.

Also, of importance is how some PLs can be biosynthesized using different nutritional pathways with implications for either optimal or pathological results. For instance, CL, which is mostly concentrated in the mitochondria and heart in particular, is optimally biosynthesized using PG from the diet, whereas oxidative stress can promote an inflammatory pathway when PG is deficient. Oxidation appears less likely in the malformation of other PLs. (52. Chen, 2018)

The fact that CL comprises a greater ratio of bodily PL content than PG implies that most PG consumption is utilized for CL. Furthermore, PG can be biosynthesized from PA, as can PI, adding another step in the process. Similarly, PE can be sourced from the diet or biosynthesized from PS, whereas PC is primarily sourced from dietary choline, but can also be biosynthesized from PE.

As will be discussed in later chapters, one of the big take-aways is that choline and PC are crucial for liver at a systemic level, while PG and cardiolipin are indispensable for mitochondria at the cellular level. In summary, PLs are essential to all the cells in the body and can be found in varying degrees in eggs, meats, liver, vegetables, nuts/seeds, dairy, grains, and citrus. (53. Christie, 2024)

Brain and Central Nervous System

The brain is undoubtedly the most complex organ of the body controlling not only conscious and subconscious activity, but via the nervous system, the vast number of bodily processes either directly or indirectly, along with life itself. Thought, memory, emotion, touch, motor skills, vision, breathing, temperature, and hunger are some of the many processes controlled or influenced by the brain.

As one of the top regulating organs, the brain is also perhaps the most downstream organ with respect to long-term metabolic consequences. Many neurodegenerative diseases such as dementia, Alzheimer's, and Parkinson's are often the end-stage result of this.

Together, the brain and spinal cord that extends from it make up the central nervous system, or CNS. The spinal cord serves as a trunk for the peripheral nervous system (PNS) which is comprised of the somatic nervous system controlling skeletal muscles, and the autonomic nervous (ANS) system which regulates organs and glands subconsciously. Together, the CNS and PNS make up the whole nervous system (NS) including the brain.

The ANS can be further subdivided into the sympathetic (SNS) and parasympathetic (PSNS) nervous systems which basically modulate the same functions in opposing ways; i.e., stress or "fight-or-flight" vs. "rest and digest".

The enteric nervous system (ENS) is a semi-autonomous network of the G.I. tract and can function somewhat independently from the brain – hence the term "second brain" – but it also bridges communication with the CNS via the vagus and splanchnic nerves for sensory input and modulation from the PSNS and SNS, respectively. Apart from local regulation of cells, hormones, and certain organs via local chemical signals and feedback; the CNS, via the PNS, primarily runs the show and sustains life. (54. ANS, 2024)

The brain is composed of three primary components – the main section divided into two lobes called the cerebrum, the cerebellum at the base of the skull, and the brainstem which connects to the spinal cord. Cerebral cortex refers to the grey matter and folds that composes the outer part of the cerebrum which contains all the neurons, whereas the inner white matter contains nerve stems that connect neurons and funnel down toward the spinal cord.

While the cerebrum controls aspects like the senses, thinking, personality traits, reasoning, spatial perception, and speech; the brainstem is responsible for more primal functions like response to the surrounding environment, coordination, facial expressions, and survival functions.

The cerebellum, located at the back of the head above the brainstem, involves voluntary movements, balance, equilibrium, and possible roles with emotion and social behavior.

The medulla in the brainstem controls certain functions such as breathing, heart rhythm, and reflex activities like sneezing and coughing. The brainstem also makes a small amount of serotonin to help regulate gut motility, circadian rhythm, and mood, but most serotonin is created in the gut. (55. Hopkins, n.d.)

Special components or glands within the brain include the thalamus, amygdala, hypothalamus, pituitary gland, hippocampus, pineal gland, and blood-brain barrier which surrounds the brain. The thalamus and amygdala together involve sensory, arousal, "fight-or-flight", emotional, and memory responses.

The hypothalamus behaves like a control center receiving signals from various parts of the brain to release appropriate signaling hormones to the pituitary gland, which in turn, dispatches various hormones to wherever needed in the body. (56. Challenged, n.d.) (57. Study Smarter, n.d.)

The hippocampus supports learning, memory, and spatial perception (space/time), while the pineal gland secretes melatonin and regulates circadian rhythms along with the sleep-wake cycle.

I know this is all a lot of information, but now given this complex, yet condensed anatomical background, we can begin exploring the nutritional implications for the brain and nervous system.

Regarding macronutrients, the majority of brain tissue is comprised of roughly 60% fat along with water and smaller amounts of protein for the nerves, hence, the widely known focus on omega-3s for neuro-protective aspects, cognitive ability, memory, and overall brain health. (58. Lifestyle Desk, n.d.)

While the actual neurons and nerve stems are made from proteins, the myelin sheath, made from omega-3 fatty acids like DHA that protect nerves, are akin to the plastic coating that shields electrical wires from short circuiting and overheating (inflammation).

Phospholipids such as PC and PE are other essential fats involved with not only signaling and cell-to-cell communication, but also the structural cell membranes of the brain tissue itself. (59. Sherzai, n.d.)

In fact, PE is the most abundant PL in the brain. Lack of PE and PC have even been linked to Alzheimer's, Parkinsons, liver disease, and other issues. Phospholipid homeostasis, critical to brain health and longevity, is regulated by both the hippocampus and the liver.

Glucose serves as the main food source for the brain, but many vitamins and minerals are crucial for different parts of the brain as well, including the blood-brain barrier. This does not necessarily imply consuming glucose directly, but that many carbohydrate foods are broken down into glucose by the body.
(60. Elizabeth Calzada, 2016) (61. Palafox-Sánchez, 2021)

The blood-brain barrier (BBB) is a thin membrane surrounding the brain that allows in nutrients, but blocks harmful substances from entering. Besides glucose and fatty acids, Vitamin C is one of the few nutrients that can cross the blood brain barrier BBB without a transporter and also has its highest tissue concentration in the brain.

This is because Vitamin C plays such an important role in so many critical aspects of the brain such as helping to metabolize glucose, countering oxidative stress, maintaining integrity of blood vessels and myelin nerve sheaths, converting dopamine to adrenaline, improving cognition, and buffering neurodegenerative diseases such as dementia and other related conditions. (62. Marano, 2018)

With the assistance of transporters, glutamine crosses the BBB to help synthesize the neurotransmitter glutamate, which can also be synthesized from glucose and vitamin B3 using a different pathway. Sodium and potassium are two minerals which can also cross the BBB freely, so the implication for all these nutrients is that they are critical for structural maintenance, cognitive function, anti-oxidation, and brain hydration. (63. Adães, 2018)

Research also shows that chronic inflammation in the body outside of the CNS may play a role in promoting Alzheimer's disease and cognitive decline in older adults along with other neurodegenerative diseases. A healthy hippocampus which supports memory and space-time cognition is critical to averting these conditions and can especially benefit from omega-3s, flavonoids, and vitamins B & D.

Another pathway for these diseases is caused by cytokines directly crossing the blood-brain barrier to trigger inflammation in the central nervous system. Either pathway can promote what's known as unregulated tau protein hyperphosphorylation and cause the breakdown of neurotransmitters in the whole nervous system leading to cognitive decline and dementia. Parkinson's and MS have also been linked to this dysfunction. (64. Keenan A. Walker, 2019)

Monounsaturated fats and omega-3s help to suppress or regulate cytokines as opposed to omeg-6 fats which tend to increase them without regulation. A healthy mediterranean or DASH diet has been suggested by experts to help manage tau dysregulation.
(65. Grimble PHD, 1998)

Glycine is an important amino acid and neurotransmitter for the NS that co-transmits along with glutamate to calm anxiety, alleviate depression, and negate mental disorders. As shown later, glycine is also essential for gut health, collagen production, bile synthesis, glutathione, and creatine. (66. Cortese K, 2023)

Certain brain elements need highlighting due to their juncture between the more instinctive and feedback signaling pathways received from the body & brain, and the more hormonal pathways sent to the rest of the body. As stated earlier, two of these components are the hypothalamus and pituitary. One of the most important functions of the hypothalamus is to link the nervous system to the endocrine system via the pituitary gland.

Whereas the hypothalamus is primarily a control center, the pituitary is more like a hormone factory, but both work together to help regulate whole-body homeostasis. Vital hormones produced by the pituitary from hypothalamus precursors are listed as follows:

ACTH – A hormone triggered by the hypothalamus hormone, CRH. ACTH stimulates primarily the release of cholesterol from the liver to form pregnenolone in the adrenals as a precursor hormone for cortisol, aldosterone, and androgen precursors for testosterone and estrogen in the adrenals and gonads.

LH and PRL – Two other hormones related to sexual development.

Dopamine – Produced primarily in the mid-brain, but also in the hypothalamus and adrenals, it is responsible for memory, pleasure, sleep, mood, learning, and systemic regulation of organs.

HGH – Human growth hormone triggered by the hypothalamus hormone, GHRH, that stimulates growth.

TSH – A thyroid stimulating hormone triggered by the hypothalamus hormone, TRH, which is a precursor to the synthesis of thyroid hormones/proteins such as T4, ferritin, and transferrin.

ADH – Regulates reabsorption and excretion of fluids in the kidneys as well as circadian rhythm.

Oxytocin – Responsible for psycho-physiological responses like social/emotional bonding and arousal.

Nutrients critical for healthy pituitary function include zinc, iron, and Vitamins D & E; with iodine, manganese, and magnesium being more involved through enzyme and regulatory feedback . Low levels of iodine and vitamin A have been shown to negatively affect communication between the pituitary and thyroid glands.

Vitamin D, B vitamins, and magnesium are also needed to mitigate and optimize depression, brain fog, concentration, memory, alertness, and improved mood respectively. As conductors of electrical signals, minerals such as calcium, iron, zinc, magnesium, selenium, and iodine are critical for neurotransmitter signaling as well as enzymatic cofactors, cellular integrity and plasticity.

Finally, for the health of BBB itself, a wide range of nutrients help to optimize and maintain its integrity. Resveratrol, curcumin, sulphoraphane from cruciferous vegetables, magnesium, omega-3s, B vitamins, and Vitamin D all contribute. (67. Schmoe, n.d.)

To recap, nutrients necessary for the central nervous system as a whole include Vitamin C, potassium, sodium, glucose, phospholipids, omega-3s, and protein, in addition to all of the nutrients listed above. Including healthy brain foods such as berries, citrus, nuts, eggs, avocados, salmon, poultry, and olive oil; while eliminating inflammatory vegetable oils and other processed foods, along with exercise proves even more beneficial to preventing not only neurodegenerative, but the whole spectrum of metabolic diseases. (68. Kubala, 2024) (69. National University, n.d.)

In summary, the brain and nervous system, with all of its parts, is responsible as the control center for a whole spectrum of processes including thought, memory, emotion, reflexes, motor skills, sensing, and all the autonomic functions to maintain whole-body homeostasis, as well as life itself. Optimal brain nutrition and a healthy liver to cleanly process nutrients that can safely cross the blood-brain barrier is therefore paramount in helping to maintain whole-body health and an optimally functioning brain.

Cardiovascular System

The cardiovascular system, which includes the heart, arteries, and veins; is the most critical vital organ system for sustaining life aside from the brain, but ironically, is also one of the least complex in terms of functionality and biochemical processes. Although complex in its nerves and tissue, the heart basically performs as a pump, sending oxygenated blood from the lungs to all the tissues of the body and returning it back to the lungs to be re-oxygenated.

Nevertheless, and aside from congenital defects, cardiovascular disease (CVD) in the form of heart attacks, strokes, and other cardio malfunctions are the leading causes of death and impairment in the U.S. and therefore, cardiovascular health has become a top priority.

As one of the most downstream organs in terms of metabolic dysfunction, maintaining vascular and neuro-muscular integrity of the heart for proper circulation is crucial, so in addition to a heart-centric diet, sustaining healthy upstream organs is paramount.

The heart is comprised of four chambers; two upper ones called atriums and two lower ones called ventricles. Contained within the right atrium is a natural pacemaker called the sino-atrial node (SA) which disburses electrical signals and cyclical rhythms throughout the heart to maintain regularity.

The brain does not control heart beats directly, but indirectly signals faster or slower rates to the SA node via the CNS. The heart also receives feedback from metabolic processes and external responses such as physical activity and emotion. (70. Ignite Healthwise, 2024)

Another interesting fact is that the heart also produces a couple of hormones called ANP and BNP. These hormones are secreted in small amounts to buffer physical stress, but increase in chronic conditions like congestive heart failure. (71. Nakagawa, 2019)

Electrolyte homeostasis is very important for cardiovascular health and also plays a key role upstream of the heart. Angiotensinogen, which is produced in the liver, is a precursor to the enzyme renin, synthesized in the kidneys, that regulates vascular constriction to increase blood pressure in response to electrolyte feedback in order to maintain balance in the system. (72. Cleveland Clinic, 2022)

Liver fibrosis, which tends to lower blood pressure through loss of filtration, increases renin as a response, but if calcified arteries do not allow vascular constriction, this may lead to excess fluid retention and electrolyte imbalances as a means of trying to maintain blood pressure. (73. Palit, 2014)

This may be why the medical field often recommends low salt for liver patients as well as heart patients even though, according to recent research and many experts, the opposite may be needed. As the saying goes, correlation does not necessarily imply causation.

Apart from the more long-term indirect effect of impaired kidney's ability to regulate electrolyte and calcium balance; diabetes mellitus (DM) and its lesser-of-the-two-evils twin, insulin resistance (IR), are some of the most destructive metabolic diseases today, having major implications for CVD and kidney disease (KD).

DM and IR have been shown to contribute to a more direct path of calcification or atherosclerosis via oxidative stress, endothelial (blood vessel lining) dysfunction, inflammatory cytokines, and impaired mineral metabolism via KD.

Fortunately, being primarily diet-related, DM and IR are largely preventable and even reversable depending on stage, but damage to the pancreas, kidneys, and heart is not quickly healed. Nonetheless, targeting these first two organs along with other upstream organs and lifestyle changes can help to mitigate damage and long-term effects of CVD. (74. Yahagi, 2016)

Of course, no discussion on CV health would be complete without noting the infamous LDL "cholesterol", but according to research, not all LDLs are created equal. While small dense LDL (sdLDL) is associated with increased CVD risk, an increase in the larger buoyant LDL (lbLDL) has been shown not to be.

This lbLDL finding can be characterized by a decrease in triglycerides alongside an increase in HDL cholesterol during tests. (75. Ivanova, 2017) (76. Froyen, 2021) (77. Berg, 2024)

Zinc may play a role in this process and has been shown to mitigate atherosclerosis and CVD. A factor that may correlate with cardiac fibrosis over time however, is a high-fat ketogenic diet, although research is needed to see if the mice study translates to humans. Research does support metabolic benefits from a healthy ketogenic diet, at least for a period of time. (78. Choi, 2018) (79. Sternberg, 2023)

Nutrition plays a critical role in cardiovascular health, along with avoiding inflammatory processed foods. Vitamin D_3 is pivotal in CV health since it plays a role in controlling atherosclerosis, but is best when paired with Vitamin K_2, which directs calcium to correct targets like bones, muscles, and teeth instead of arterial walls.

Furthermore, the role sunlight plays in producing Vitamin D may be important in reducing excess cholesterol since Vitamin D is biosynthesized from UVB radiation in the skin using cholesterol as a substrate. (80. Surdu AM, 2021) (81. Sunlight Institute, 2017)

CoQ10 is a coenzyme nutrient essential for promoting energy production in cell mitochondria (ATP), but is especially concentrated in the heart and relies on tyrosine and nutrients like vitamins C, E, B12, B6, B5, and selenium in order to biosynthesize in the body. It can also be taken as a supplement or found in meats, organs, fish, poultry, dairy, and legumes. (82. Cytoplan, 2017)

Taurine is a powerful amino acid that in addition to digestive and other benefits, also helps to support mitochondria, cardiovascular health, and even heart failure by increasing nitric oxide and ATP. Other important amino acids for the heart include branch-chained aminos (BCAAs) leucine, isoleucine, and valine; arginine, carnitine, and glycine. (83. Jong CJ, 2021) (84. Amino Science, 2018)

Optimum CV health is supported by a variety of nutrients besides ones listed above such as magnesium, omega-3s, monounsaturated fat, phospholipids, polyphenols, quercetin, B9, B12, and resveratrol with CoQ10 and berberine being the most potent supplements. Heart healthy foods include fish, olive oil, leafy greens, berries, avocados, whole grains, beans, nuts, garlic, tomatoes, dark chocolate, and chia & flax seeds. (85. Qardio, 2017) (86. AJmera, 2025)

In summary, the cardiovascular system is vital to life and can take a beating, but lifestyle choices should be considered from an earlier adult age in light of the long-term and detrimental effects that can accumulate over time. Maintaining healthy upstream organs like the liver, pancreas, and kidneys is crucial for proper nutrient absorption, blood vessel health, and reducing calcification.

Respiratory System

As the second most vital organ in terms of criticality, the respiratory system is vitally responsible for absorbing oxygen, expelling carbon dioxide, and immune support, but tends to be more downstream regarding metabolic consequences than other organs.

The respiratory system consists of the trachea (windpipe), bronchial tubes, and the lungs which are connected to them. Bronchioles, which are smaller branches that connect the bronchial tubes to the lungs, terminate at the alveoli or air sacs in the lungs where the exchange of oxygen and carbon dioxide occur via capillaries in the alveoli. Cillia are fine hairs in the airways that work to filter out dust and other irritants. External air breathed in is moisturized and warmed to body temperature before reaching the air sacs.

In addition to taking in oxygen and expelling carbon dioxide, the lungs also enable functions like speech and smell. As part of the innate immune system, epithelial cells lining the alveoli air sacs protect against bacteria, viruses, allergies, and other harmful substances. Not commonly known is that lung health is also connected to the liver and kidneys. (87. Cleveland Clinic, 2024)

The renin-angiotensin system (RAS), more often associated with how the kidneys regulate fluid balance and blood pressure, is involved with lung function as well. Together with renin from the kidneys, the liver produces angiotensin I which is sent to the lungs to synthesize angiotensin II in the capillaries of the lung with the aid of angiotensin-converting enzyme (ACE).

Damage in the liver and/or kidneys can inhibit this pathway and therefore, disruption in the RAS which has been implicated in lung diseases such as COPD and pulmonary fibrosis among other causes. (88. Kuba K, 2006) (89. Taskin E, 2017)

One of the most vital aspects of lung function is supporting healthy maintenance of the surfactant tissue that lines the alveoli air sacs. Some of the most important nutrients for surfactant lung health are phospholipids. PL concentration and type vary throughout the body, but the most abundant found in lung tissue are phosphatidylcholine (PC) and Phosphatidylglycerol (PG). PC accounts for 80%, and although about only 1% in the body at large, PG accounts for 7-15% in the lungs suggesting a very critical role there as well. (90. Agassandian M, 2013)

In the lungs, PL-enriched surfactants maintain fluid-free air sacs and optimize gas exchange of oxygen and carbon dioxide. PC is found in eggs, poultry, fish, and beef, while PG is found in spinach, soy, oats, eggs, buckwheat, chickpeas and barley, but can also be synthesized from PA in cabbage. Monounsaturated fats and vitamin E from foods like avocados and olive oil help reduce inflammation.

Research shows that other important factors affecting and regulating surfactant production include cortisol, thyroid hormone (T3), and growth factors. In fact, promising evidence even shows that T3 can inhibit mitochondrial dysfunction in pulmonary fibrosis. (91. Alves E, 2018) (92. Breitzig MT, 2018)

As a supplement, N-Acetylcysteine (NAC) is somewhat effective, but its upstream relative GLRX (precursor to glutathione) has shown more promise in trials. Glutathione itself is produced in, and is protective of the lungs. (93. Anathy V, 2018) (94. Tuzova M, 2014)

In addition to phospholipids, lung health is sustained from a host of nutrients. Among these are zinc, Vitamin C, vitamins A & D, Resveratrol, Vitamin E, selenium, omega-3, magnesium, and curcumin. The most potent of these regarding lung disease appear to be zinc, Vitamins A, C, and D; and resveratrol, but all the others are important for lung support and alleviation of conditions like asthma and general lung health.

It was surprising to note that according to research, zinc may actually be able to reverse pulmonary fibrosis. Vitamin A is shown to help regenerate lung tissue and inhibit the risk of emphysema as does Vitamin D also. Vitamin C and resveratrol protect against pneumonia and COPD as well. (95. Landry, 2025)

In summary, the respiratory system, and particularly the lungs, are critical to life and sustaining health. Next to the brain and heart, the lungs may be one of the most downstream organs in terms of long-term cascading metabolic imbalances, but can also be affected by more near-term immunol and genetic consequences, stressing the need for proper nutrition.

In addition to the foundational aspects of phospholipids, the lungs can heal and even thrive with optimal nutrition from a variety of sources. It has also been shown that hormones play a critical role, highlighting a dual path not only directly from the adrenals and thyroid, but also via the kidneys from upstream organs such as the liver, which in turn, helps to regulate the adrenal and thyroid glands.

Kidneys and Implications from Liver Health

As one of the two vital organs responsible for detoxification, the kidneys are small organs that filter the blood, remove wastes, are involved with blood pressure, and regulate electrolytes and fluid balances. In addition, several conditionally-essential amino acids like glycine, ornithine, and arginine are biosynthesized in the kidneys whereas tyrosine, cysteine and taurine are regulated by the kidneys. (96. J Clin Invest, 1980)

Unlike the liver, the kidneys are also end-stage processors at the crossroads of several complex and iterative feedback loops and not as resilient in healing, so balancing the proper functional medicine approach for those with chronic kidney disease (CKD) may be a challenge. Although regulators themselves, kidneys are affected downstream of the liver, and subsequently, the thyroid and adrenal glands which can affect electrolytes, aldosterone and filtration rates.

Research also shows that non-alcoholic liver disease (NAFLD) is a high-risk factor for CKD since an impaired liver can release pro-inflammatory cytokines into the bloodstream which filter through the kidneys. Moreover, an unhealthy liver and spleen can increase the detox load on the kidneys. (97. Friedman, 2003) (98. Umbro, 2021)

Restoring health to these upstream organs by meeting their respective nutritional needs is therefore a crucial starting point for kidney health. Proper supply and regulation of structural proteins and PLs that comprise kidney tissues is foundational as well.

The phospholipids PE and PC are the main constituents of the kidney's cellular membranes, but oxidative damage can be a risk factor when they are infused with oxidized polyunsaturated fats (PUFAS), so vegetable oils should be avoided. (99. Florens N, 2016)

Many in the medical field believe kidney damage is irreversible, but recent research suggests kidneys can be healed naturally through three pathways – TUDCA (a powerful bile acid), intermittent fasting, and a healthy ketogenic or mediterranean diet. Whereas chronic kidney disease (CKD) caused by lifestyle may be more responsive to TUDCA, polycystic kidney disease (PKD) may only be responsive to the latter two. (100. Yoon YM, 2019)

Research has shown that kidney cysts in PKD can shrink when deprived of glucose in lieu of ketones produced during fasting. The keto diet, espousing moderate protein and high fat, may be tweaked to allow only moderate healthy fats and more cruciferous vegetables along with polyphenol and resveratrol foods like red grapes, berries, and pistachios. (101. Fernandez, 2019) (102. Den Hartogh D, 2019)

Conflicting opinions about certain electrolytes such as potassium with CKD could suggest that the benefits vs risks may depend on how far along the spectrum of recovery a person is. In other words, the less advanced the stage, the more resilience to higher or more normal amounts consumed. Perhaps a modified intake of salt, magnesium, potassium and vitamin D in their recommended ratios depending on disease stage is a good approach. (103. Picard K, 2020)

Also, recent research has shown very promising benefits of resveratrol in helping to heal the kidneys without having to balance the pros and cons that come with some of the other nutrients. Along with resveratrol; vitamin D, exercise, and a Mediterranean diet have been shown to upregulate the klotho gene, a path responsible for improving not only kidney health and longevity, but brain as well. (104. Chu SH, 2021) (105. Buchanan S, 2020) (106. Wu S, 2022)

Including monounsaturated fats (MUFAS) like olive oil, avocados, and nuts along with consuming enough salt to upregulate magnesium and Vitamin D for calcium homeostasis could be key as well. A healthy mediterranean/keto-light diet with sufficient electrolytes modified as needed along with exercise seems to be the best approach. Root vegetables, resveratrol, and berries are some of the best sources of nutrients along with vitamins A, B, C, D, E, & K, calcium, iron, potassium, and manganese; in addition to high amounts of antioxidants for healthy cells. (multiple sources)

In summary, the kidneys, like other organs, may be restored, but the timeframe is probably quite a bit longer as opposed to the liver. As a regulator of adrenal and thyroid glands, which in turn help regulate the kidneys, the liver not only has a downstream effect on the kidneys via this pathway, but a healthy liver can also directly relieve some of the detoxification load on kidneys by functioning more resiliently; thus, healthy liver foods should be considered also.

Pancreas

The pancreas is an organ that serves a dual function as part of both the digestive system, with its role of releasing digestive enzymes from the exocrine gland, and endocrine gland with its role in releasing hormones like insulin and glucagon. The main enzymes released by the pancreas are lipase, protease, and amylase for fat, protein, and carbohydrate digestion, respectively, along with release of bicarbonate to neutralize stomach acid in the duodenum.
(107. Cleveland Clinic, 2024) (108. PanCan, n.d.)

Insulin and glucagon work opposingly to either decrease or increase blood sugar correspondingly. Somatostatin, produced in both the pancreas and hypothalamus, acts as a feedback hub inhibiting these hormones and others such as CCK in the duodenum, growth hormone from the pituitary, and thyroid hormone, but is likewise regulated by these same hormones in a feedback loop.

In addition, somatostatin is also regulated by the vagus nerve of the parasympathetic nervous system, so maintaining a healthy vagus nerve is important for promoting pancreatic homeostasis.
(109. You and Your Hormones, 2021) (110. Babic T, 2016)

At the crossroads of many chemical and neural feedback signals, this stealthily complex organ is vitally important to both the digestive and endocrine systems, but thanks to modern medicine, living without one has become possible with an adapted lifestyle.
(111. Abramowitz, 2016)

Preferably this would be avoided by healing the pancreas naturally, and current research suggests a path may exist using GABA to help restore insulin beta cells. Restoring fibrotic exocrine cells, though is uncertain, but research confirms that zinc helps with other organs.

Apart from supplementation, GABA is biosynthesized from glutamine, but also needs Vitamin B6 as a cofactor to help facilitate the process. Besides diabetes and CVD related issues, the pancreas is also at risk for diseases such as pancreatitis, fatty pancreas, and pancreatic cancer. Amazingly, berberine has been shown to mitigate these issues by stimulating AMPK and apoptosis, or programmed cell death for cancer cells. (112. Rutter, 2017) (113. Vlavcheski F, 2022)

Non-alcoholic fatty liver disease (NAFLD) is a major risk factor for non-alcoholic fatty pancreas disease (NAFPD) with a correlation of about 70%. One study shows that only one quarter of pancreatic patients had normal livers, and since a large portion of the pancreas is digestive related, this suggests the downstream effect of NAFLD is highly correlated with the exocrine part of the pancreas, whereas diabetes is unrelatedly associated with other metabolic pathways affecting the endocrine part of the pancreas. (114. Zhang CL, 2021)

In other words, one may have enzyme insufficiency resulting from both a fatty liver and pancreas without major blood sugar issues, but conversely, one may have insulin related damage without a fatty liver through insulin resistance in the muscles or hyper gluconeogenesis, breaking down protein and fat into glucose. (115. Ulasoglu C, 2021) (116. Spero, 2013)

In summary, the pancreas is a mighty bi-functional organ that helps to facilitate healthy digestion, regulate blood sugar levels, and assist with total body homeostasis, so proper nutrition is key. Nutrients essential for the pancreas include Vitamins C & K; leafy green, berry, and olive oil polyphenols; cysteine, cruciferous flavonoids, and probiotics.

Cysteine can be synthesized in the body and/or by consuming meat, eggs, fish, dairy, nuts, and legumes from pasture fed, wild, and organic sources. As shown later, many of these whole, balanced foods overlap those necessary for healthy gut and liver function as well. (117. Liver Doctor, n.d.) (118. Juechter, 2022)

Spleen and Lymphatic System

The spleen is a small blood-filtering lymphatic organ located next to the stomach. While it is not directly connected to the lymphatic drainage system, the spleen filters blood from the whole body, including any residual waste left in lymph fluid that merges with blood circulation at the end of its course. Long known for its ability to recycle old blood cells and degrade pathogens, the spleen also removes cellular waste products, stores blood, and generates white blood cells and antibodies. (119. Cleveland Clinic, 2021)

Although the liver can compensate for loss of the spleen, the spleen supports the liver by pre-filtering some of the waste that would otherwise need to be degraded by the liver. Conversely, a damaged liver can stress the spleen and cause it to enlarge as a result.

Current research suggests that assessing the status of the spleen may become a non-invasive option to invasive tissue testing for liver and pancreatic disease. Indeed, ancient Chinese medicine has long recognized the spleen as a key organ supporting digestive organs, albeit more indirectly. (120. Tarantino G, 2013) (121. Wu, 1998)

Other components of the lymphatic system are bone marrow, thymus, lymph nodes, and the network of lymph vessels that span from head to foot. Like the spleen, bone marrow also makes white blood cells, along with red blood cells. The thymus is responsible for producing immune T-cells which migrate primarily to the bloodstream to destroy any systemic pathogens and cancer cells, but also store within lymph nodes as a backup.

Lymph nodes destroy pathogens and arrest cancer cells as lymph fluid passes through lymph vessels into blood circulation where T-cells await to do their job, after which, the spleen cleans up the aftermath. (122. Cleveland Clinic, 2023) (123. Gardner, 2022)

As the organ that trains up nascent white blood cells into trained killers, the thymus has caught renewed interest as immunological research reveals therapeutical pathways for thymus regeneration molecules IL22 and BMP4. The thymus tends to shrink and atrophy with age, explaining the immune vulnerability for older adults, but maintaining healthy adrenals assists in rapid response to the immune system when threats are detected. (124. Keown, 2018) (125. Deak, 2008) (126. Stanford, University Medical Center, 2012)

A variety of nutrients known to support the thymus include vitamins A, C, D, and E, along with iron, zinc, and selenium. To summarize, although the spleen and lymphatic system is often overshadowed with regards to the main energetic organs in the body, they are an indispensable part of the system contributing to health and longevity. (127. Gardner, 2022)

Thyroid Function and Implications from Liver Health

Known as the body's thermostat, the thyroid gland plays a very important role regulating many of the body's functions including temperature, metabolism, energy production, and kidney function. PTH, is an important hormone, that while it comes from the parathyroid glands, not thyroid, assists the kidneys with balancing electrolytes and calcium homeostasis. T3 is primarily a liver hormone that circulates throughout the body to regulate these processes, but is also tied to healthy lipoprotein (LDL) homeostasis.

The hypothalamus gland produces and sends a hormone called TRH to the pituitary gland which makes TSH for the thyroid. The thyroid, enriched by tyrosine, iodine, iron, and copper, then signals TSH to make the hormone T4. Selenium, iron, and zinc are critical for the last step of converting T4 to T3 for healthy thyroid function.

Deficiencies in B2, B6, B9, B12, vitamin A, and vitamin D impair thyroid function. Since the liver produces most of the body's T3, liver damage and/or deficiency in selenium, iron, and zinc reduces T3 synthesis. (128. Krishnamurthy HK, 2021) (129. Childs, 2025)

TSH from the pituitary is also a precursor to ferritin and transferrin, which are proteins produced mainly in the liver to regulate intracellular iron stores in non-toxic form to prevent either deficiency or overload in the blood iron levels. Transferrin is transported to cells throughout the body, whereas ferritin is concentrated mostly in the liver as an iron storage backup. Ferritin is also concentrated in small amounts in the spleen, bone marrow, duodenum, skeletal muscle.

Although iodine deficiency most often gets the spotlight regarding thyroid problems, in many cases, hypothyroidism is likely caused by either impaired T3 production or insufficient ferritin production leading to hypothyroid-mimicking iron deficiency in the blood, both cases of which may result from an impaired liver. In summary, though many nutrients are involved in its homeostasis, thyroid function is particularly sensitive to iron regulation as are the hypothalamus and liver health as well. (130. Hedberg, n.d.)

Adrenal function, hormones, and implications for liver health

The adrenal glands are powerful organs sitting atop the kidneys that are responsible for regulating several hormones for metabolism, homeostasis, reproduction, and acute survival. They also sit at the juncture between emotional responses and how the body reacts physiologically downstream. Adrenals can have a profound effect on the liver, as mentioned in Part 1, from elevated cortisol via chronically high stress or trauma.

The primary hormones produced are adrenaline, aldosterone, cortisol, and androgens (sex hormone precursors). All the hormones are activated by the pituitary gland hormone ACTH, which in turn is activated by CRH from the hypothalamus, but only the last three are synthesized from cholesterol as a building block; whereas adrenaline is made from tyrosine, which in turn can be biosynthesized from phenylalanine. The main hormones produced in the adrenals are as follows:

Adrenaline – The hormone famously associated with "fight or flight" response which increases heart rate, blood pressure, and more relaxed breathing in times of stress. Unlike the other hormones, synthesis for adrenaline begins with phenylalanine being metabolized in the liver for conversion to tyrosine, which is a precursor for dopamine, which in turn is a precursor to noradrenaline in the final step to become adrenaline.

High levels of tyrosine have been connected with advanced liver disease, but not the cause of it. As with many systems, an alternate pathway exists for adrenaline. As a backup, noradrenaline performs the same function, but originates from neurons in the sympathetic nervous system in addition to the adrenals.

Pregnenolone – Primary non-adrenaline precursor hormone for producing other hormones such as aldosterone, cortisol, and androgen sex hormone precursors like DHEA and progesterone.

Aldosterone – As previously mentioned, involved with regulating electrolytes, blood pressure, and kidney function. Although high aldosterone levels have been linked with certain metabolic diseases, it appears more likely that they are a correlation in response to a system out of balance rather than a cause of it. Aldosterone is synthesized from cholesterol, which in turn is regulated by the liver.

Cortisol – Responsible for regulating metabolism and immune system. In the absence of carbohydrates, cortisol will rise in order to metabolize protein and fats to maintain balance in glucose levels, so while beneficial for short term health conditions, sustained low-carb diets are probably not good in the long term for most people. Like aldosterone, cortisol is also biosynthesized from cholesterol.

Androgens – Primarily DHEA and progesterone also synthesized from cholesterol, which are precursors to testosterone and estrogen produced in the gonads, but also by the adrenals in smaller amounts; however, the adrenals compensate more in women after menopause.

In summary, the adrenals perform a large number of functions from regulating metabolism and system homeostasis, to survival and reproduction. Unlike other hormones such as aldosterone and cortisol, adrenaline as well as androgens don't appear to be dependent on any direct homeostasis feedback loops such as electrolyte or iron homeostasis, but are still regulated upstream by the liver via cholesterol (androgens) and phenylalanine (adrenaline).

However, since adrenaline has a backup pathway via brain neurons and dietary tyrosine instead of phenylalanine conversion, it can function fairly independent of liver health, which makes sense from its role as a survival hormone. Aldosterone doesn't have that option.

The adrenals are nourished from a wide variety of nutrients starting with vitamins C, B5, B6, & D, magnesium, choline, zinc, and healthy fats including omega 3's from foods like fish avocados, olive oil, nuts, olives, eggs, butter, and fish.
(131. Wikipedia, n.d.) (132. Axe, 2023)

Gastro-intestinal Tract

As stated earlier, the G.I. tract is where it all starts with regards to consuming and absorbing nutrition. The journey begins with the salivary glands signaling the release of enzymes in the mouth and pancreas for carbohydrate and fat digestion upon smelling and tasting food. In the stomach, gastric acids along with pepsin break down proteins into amino acids.

Once entering the small intestine, where the vast majority of nutrients are absorbed, bile from the gallbladder does the heavy lifting of digesting fats, but fat, carbohydrate, and protein enzymes from the pancreas help to complete the job. A small amount of enzymes is also released in the upper small intestine itself. Next, bacteria and microbes assist in breaking down particles through fermentation of undigestible fiber in the large intestine to create small chain fatty acids (SCFAs) such as butyrate.

These are consumed and used by the intestinal lining to maintain a healthy gut barrier for proper absorption of nutrients, prevent gut inflammation, and ward off problems from leaky gut.

One of the crucial aspects of healthy digestion is how the body manages changing pH levels throughout the various stages in the G.I. tract. Once food in the stomach comes into contact with HCL having a pH of 0.8, the resulting breakdown mixture called chyme exits the stomach through the pyloric sphincter into the duodenum with a pH between 1 and 3. (133. Sensoy, 2021)

This triggers duodenal secretion of mucous and the hormones secretin and cholecystokinin (CCK) for activation of bicarbonate & pancreatic enzymes, and bile respectively. Bicarbonate from the pancreas along with bile from the gallbladder reduces the acidity and pH to around 7, while bile also, along with enzymes, furthers the digestion process and begins the extraction of nutrients.

All the vitamins, most minerals, amino acids, fatty acids, and glucose formed from ingestion are absorbed through the small intestine, whereas salt, potassium, water, and endogenously created fat-soluble and B vitamins (B2 and B5) are absorbed in the colon.

Once completely digested, water-soluble vitamins B & C, glucose, and amino acids are absorbed through the small intestinal lining directly into the bloodstream for transport to tissue cells; whereas, fatty acids and fat-soluble vitamins A, D, E, and K are sent to the liver to be processed and/or stored before entering the bloodstream.

A certain amount of glucose is also sent to the liver to be stored as glycogen for later use. Along with Omega-3s, amino acids, and certain vitamins, glucose is particularly important as an energy source for the brain, so proper uptake is essential. (134. Kiela PR, 2016) (135. Sprouse, n.d.)

Even sunlight has been indirectly linked to gut health. As a precursor to endogenous Vitamin D via cholesterol as a substrate, sunlight has been shown to be effective in digestive health, and by direct implication, may also help regulate cholesterol metabolism. In addition to increasing the diversity of gut microbiota, sunlight is also responsible for stimulating the only endocrine hormone produced in the stomach, serotonin. In fact, up to 95% of serotonin is made in the stomach, not the brain.

In addition to regulating circadian rhythm, mood, and gut motility, serotonin also regulates sex functions, blood clotting, and wound healing. Besides sunlight, tryptophan aided by vitamins B6, B12, and folate are the primary precursors to serotonin production. Notwithstanding all these nutrients, other major influences on gut health exist also. (136. GrassrootsHealth, 2021) (137. Scaccia, 2024)

A dysfunction with the gallbladder and/or bile via the liver, along with impaired bicarbonate secretion into the duodenum can alter its pH and damage the villi known as villous atrophy. Not only can this cause malabsorption, but if bile reverses back through the pyloric sphincter into the stomach, it can cause bile reflux, or gastritis.

Moreover, harmful gut microbes from a bad diet over time can alter the very bile acids responsible for suppressing their assault on gut integrity. The resulting malabsorption and inflammation disrupt nutrient absorption and biosynthesis of reabsorbed bile acids leading to an even more impaired liver, which in turn, secretes unhealthy bile in a vicious feedback loop. (138. Larabi AB, 2023)

To complicate things worse, if obstructive gallstones or bile sludge in the common duct occur, this can signal not only liver damage, but also pancreatic disorder, blocking of enzymes, and reduced secretion of CCK and its mediating effects. (139. Wang HH, 2019)

Another parallel pathway for gut dysbiosis arises from the brainstem and spinal cord of the autonomic nervous system (ANS) via the vagus and splanchnic nerves, respectively. This neurological disruption, which affects gut motility, neurotransmitter secretions, and blood flow, can originate from either stress via the ANS or through impairment to the hypothalamus and brainstem from bile acids leaked into blood circulation via liver disease.
(140. Yeo XY, 2023) (141. Harvard Health, 2019)

Not everyone with unhealthy livers and G.I. tracts experiences acute symptoms, but if unchecked, these issues can lead to long-term metabolic diseases in a cascading domino effect as cited in earlier sections. These scenarios illustrate the complex interdependency between the liver, gut, and brain and how important it is to follow a proper diet and alleviate chronic stress and toxic emotions.

A proper study on gut health would not be complete without prebiotics, probiotics, and other good bacteria which help modulate the gut biota-bile acid feedback cycle affecting the gut, liver, and brain. As noted in the beginning, gut bacteria communicate with the brain, forming neurotransmitters like serotonin, dopamine, and GABA to regulate mood, emotions, sleep, and motivation.

Beneficial bacteria also help digest food and fiber to produce vitamins, fatty acids, and anti-inflammatory compounds, in addition to influencing bile acids, so foods like fermented dairy and vegetables, mushrooms, a wide-variety balanced diet, and probiotics as needed are key to easing gut dysbiosis and improving gut biome.

Reducing sugar and other processed foods along with intermittent fasting (as necessary) are also important because they diminish the competition of bad bacteria in the gut while allowing a head start for good bacteria to proliferate more.

Since life itself depends on the absorption of all these nutrients, it is vital to maintain healthy enterocyte as well as the more general epithelial cells of the intestinal lining to promote overall health. Because the gut lining is repaired directly by ingested nutrition, applying a healthy diet is key to healing the G.I. tract.

Glutamine is widely thought to be the most important amino acid for the small intestine, followed by glycine, but as shown earlier, PE is necessary to maintain cellular membranes while other less-known attributable nutrients are as needed as well.

Important to note is that glycine is the most common amino acid substrate for the formation of collagen, which is well known for not only gut health, but also building blocks for tendons, ligaments, skin, hair, and nails. (142. Axe, 2024)

Vitamin A is rarely discussed in nutritional circles for gut health, but is crucial for mucosal and epithelial cell growth in addition to regulating iron uptake. Vitamins B2, C, and D, along with zinc, magnesium, and potassium, are also beneficial to overall intestinal homeostasis. Finally, bromelain extracted from pineapple, is known for helping to protect the gut lining as an anti-inflammatory agent aiding digestion. (143. Rao JN, 2010) (144. Griffen, 2024)

In summary, we may get away with poor diets in our youth, but sustaining a healthy gut and liver as we age is essential for maintaining the vibrant health of the liver/gut/brain axis as well as system-wide homeostasis, preventing autoimmune disease, and averting a slow-motion cascade of metabolic disorders.

Liver and Gallbladder

By now, it is apparent just how important the liver is to overall health. As a core focus for this book, the liver has been shown to be the most important organ for regulating whole-body metabolic balance. Not only is it designed as the only organ able to re-grow its own amputated tissue, but as the master organ, it is the most relatively autonomous organ from the brain and responsible for over 500 functions.

Listed below are some of the many important roles the liver plays in metabolism:

1. Receives/disburses absorbed nutrients and filters blood from digestive organs such as the stomach, intestines, gallbladder, and pancreas, along with degraded blood cells and pathogens from the spleen.
2. Primary detoxification organ that degrades toxic substances down to harmless water-soluble particles to be eliminated.
3. Produces and recycles bile for digestion in a feedback cycle which includes intestinal re-absorption.
4. Supplies the gallbladder with bile to store for secretion during meals.
5. Produces or regulates hormones like hepcidin for iron homeostasis, angiotensinogen for the lungs & adrenals, T3 for thyroid, aldosterone for the kidneys, and many of the adrenal hormones.
6. Organ where most non-essential and conditionally-essential amino acids and proteins are biosynthesized or catabolized.
7. Produces plasma proteins like transferrin, albumin, and others for cellular/fluid control and coagulation.
8. Processes and regulates cholesterol and glucose storage.
9. Stores and regulates vitamins and minerals.
10. Responsible for overall homeostasis in the body.

The liver is so important, it receives two blood supplies; one from the portal vein which consolidates the blood supplies listed above, and one from the hepatic artery which brings in a fresh supply from the heart to support liver tissues. When the liver is impaired and not functioning properly, the whole body can be affected by a myriad of different syndromes, but when it is healthy, the many interdependent systems and complex feedback loops are kept in balance, especially the liver-gut-brain axis. (145., 2008)

An unhealthy liver can lead to an impaired gut and brain, but conversely, an unhealthy gut can lead to an impaired liver and brain from an unhealthy gut biome, so retaining optimal absorption and feedback regulation between the gut and liver is essential.

In normal physiology, about 95% of bile flowing into the duodenum is recycled back to the liver with only 5% being eliminated. While new bile is produced in the liver, existing bile acids separated in the gut into secondary bile acids to degrade toxins and waste products, are re-absorbed through the gut and back to the liver to be ionized and combined with taurine and glycine to form bile salts.

Farnesoid X receptors (FXR) in the nucleus of hepatic cells are responsible for regulating bile acid homeostasis and transport out of the cells. At the crossroads of contact, the transporter ABCB4 moves phosphatidylcholine (PC) from inside hepatic cell membranes into bile capillaries of the liver for extraction by taurine conjugated bile salts passing through these channels.

Since PC helps neutralize harsh effects of the incoming bile salt/cholesterol mixture, properly activated ABCB4 is essential for bile/cholesterol homeostasis in the liver, while taurine as well as glycine play a critical role. Bile duct and gallbladder health is also greatly affected by the quality of bile produced in the liver.
(146. The Medical Biochemistry Page, 2025) (147. Sundaram, 2008)

Tauroursodeoxycholic acid (TUDCA), a powerful bile acid originally derived from bear bile, is now made synthetically and reportedly has great benefits in helping to heal liver and gallbladder disease, or at least improving conditions resulting from dietary consequences and suppressed negative emotions. Humans only produce a small amount of TUDCA, so it needs to be supplemented if needed.

Those who don't respond to TUDCA (or even if they do) may have a genetic mutation to the ABCB4 gene which can lead to reduced bile synthesis in a condition called Low phospholipid-associated cholelithiasis or LPAC, where OCA has been shown to help with those not responding to UDCA or TUDCA. In situations like this, another nutritional pathway may be exploited also. (148. Soret, 2023)

Of several bile acids within the system, chenodeoxycholic acid (CDCA) is the most effective activator of the FXR receptor and has been used similarly to TUDCA. CDCA is biosynthesized mainly from taurine and to a lesser extent, glycine. Taurine can in turn be biosynthesized from cysteine and methionine, so a lack of these nutrients can contribute to bile inefficiency. (149. Sepe V, 2016)

As a later-stage component of bile formation, PC is biosynthesized from PE, but can also be acquired through the diet, so PC and choline are other nutritional pathways that should be pursued in all cases since they are known to be great for healing the liver, reducing fat, and regulating triglycerides in the liver.

Although fairly autonomous, the liver does receive feedback regulation from what's known as the hypothalamus-pituitary-adrenal (HPA) axis, mainly with regards to stress signals and circadian rhythm. In what could be termed the HPA-Liver or Liver-HPA feedback cycle, liver dysfunction could begin with either the liver through genetics or poor diet, or from chronically elevated adrenal stress hormones like cortisol and adrenaline via the HPA pathway as described earlier in the <u>Proactive Steps</u> section.

Regardless of origin, congested bile acids in a damaged liver can leak into systemic circulation and accumulate in the hypothalamus which signals further disruption to the HPA axis and cholesterol/ bile acid biosynthesis in the liver. This is one example how stress and diet can affect the liver-brain axis. (150. Petrescu, 2018)

Along with ample amounts of taurine, glycine, and PC or choline, nutrients essential for and stored in large amounts by the liver include glycogen, cholesterol, triglycerides, vitamins A, D, E, K, B12, folate, and minerals like copper and iron. Of these, B12, folate, and iron, along with magnesium and zinc are crucial nutrients for DNA synthesis and repair. (151. Fenech, 2020)

The liver stores only small amounts of B1, lasting up to two or three weeks, so since B1 is needed almost on a daily basis, an impaired liver makes retaining enough of it even more challenging without supplementation. Although not stored, B5 is also necessary to assist in metabolizing fats and detoxification, while polyphenols are vital for helping to mitigate inflammation, toxins, and fibrosis.

Finally, essential phospholipids like PC and PE needed for cell membranes and signaling pathways by the brain and whole body, are both needed and regulated by the liver, so maintaining a healthy liver is paramount.

To summarize, the liver is an amazing organ that promotes systemic balance and can take an incredible amount of abuse, but still has its limits. Aside from inherited defects or very late-stage disease, the liver can almost certainly be healed by eliminating toxins, vegetable oils & high fructose, consuming optimal nutrition, and detoxifying chronic adrenal stress from old emotional wounds.

Senescent Cells and Low-Grade Chronic Inflammation

Up till now, this study has highlighted the dysfunctional and deficient nature regarding various organ systems and nutrients; however, there is another, more universal malady at play in the body known as low-grade chronic inflammation (LGCI). LGCI is a common characteristic of many diseases, such as obesity, type 2 diabetes, CVD, chronic respiratory disease, kidney disease, cancer, and the like. (152. D'Esposito, 2022)

LGCI is subclinical in nature (harder to detect) and recognized as a distinct type of inflammation as opposed to acute inflammation from more specific ailments and metabolic diseases. LGCI is triggered by immune cells to help restore healthy tissue but can also become chronic by promoting proinflammatory cytokines if not resolved through lifestyle changes to improve conditions like obesity, tobacco use, physical inactivity, stress, and unhealthy diets.

One pathway for LGCI may result from cellular senescence which occurs when damaged cells don't repair and multiply, nor do they die off and recycle, but simply linger around to damage nearby cells by triggering inflammation. This process is seen in aging, neurodegenerative, and many life-span decreasing metabolic diseases.

In addition to the lifestyle changes cited above, consuming a healthy mediterranean-like diet and removing inflammatory foods like vegetable oils, white flour, high fructose, excess sugar, and other processed compounds can change the whole direction of health and longevity if there is a commitment and desire for lasting change.

Whether from a specific ailment or system-wide inflammation, having an understanding of nutrients and how all the body functions work together is key to formulating a path forward. By looking under the hood, we not only reveal what needs to be addressed, but also gain an appreciation of the beauty and complexity of the most highly engineered system in the universe.

Part 3 – Select Macros, Micros, and Other Key Nutrients

In Part 2 of this study, we covered a more in-depth, but condensed analysis of organ systems, their nutritional pathways, and the downstream effects of how poor dietary choices and stress can lead to cascading metabolic consequences over time.

Although many nutrients were listed throughout those sections, this portion of the book will zero in on several key nutrients that either serve at critical junctures for total-body homeostasis or just appear to be fantastic nutrients across multiple pathways for systemic health. Hence, for the purposes of trying to convey a more targeted approach with regards to maintenance of upstream health, Part 3 will focus on just a few of these key aspects.

Protein – Not Just for Bodybuilders and the Young

As with many rebuffed nutritional theories, it turns out we need at least as much, if not more protein as we age, not less, due to decreased enzymatic efficiency and gut absorption. Protein in the form of lean muscle mass helps the body become more resilient to fluctuations in metabolism through increased storage capacity.

Because of the many amino acids and their various functions, from building muscle and bone to help store protein and nutrients, to producing neurotransmitters and hormones, to assembling other tissues in the body, and to bolstering the immune system; ample dietary protein is absolutely necessary for lifelong vitality.

Current research cites that per body weight, 1.0 g/kg - 1.6 g/kg (0.45 g/lb.- 0.73g/lb.) per day of protein is required depending on age and health goals. The low end covers young and healthy older adults with lower physical activity, while the high end covers young athletes and older frail adults or those who perform weight training. (153. Campbell WW, 2023)

Apparently, the idea of kidney damage from high protein diets in healthy people is outdated according to recent research, although those with kidney disease are advised by kidney health organizations to simply consume the normal recommended amounts, not less.

The overlap of various proteins and amino acids in a well-balanced diet reveals not only the brilliant engineering, but also gracious design of our Creator in providing a wide variety of foods for a wide diversity of climates, cultures, and situations. Looking deeper however, illustrates multiple nutritional pathways for obtaining necessary proteins and amino acids through both diet and endogenous production of end-products aminos in the body through biosynthesis using various enzymatic reactions.

The trade-off appears to be more energy efficient consumption at the expense of fewer metabolized benefits along the way, versus gaining added chain-of-process nutrients at the expense of more energy required from biosynthesis in the body. Thankfully, from consuming a diverse diet, we can achieve a balance of both and let the body decide which paths to prioritize. The following section explains this concept with AAs, but it applies to other nutrients also.

Intriguing Roles of "Conditionally Essential" Amino Acids

Amino acids (AA), both essential and non-essential, are necessary and important to the body for multiple functions ranging from structural tissues and hormones to nerve generation and immune function, but throughout this research, several amino acids have resurfaced repeatedly.

Depending on how bioavailable they are in the diet, some of these "conditionally essential" amino acids can also be considered non-essential, but only if the body is running optimally and able to convert them from precursor AAs properly. As noted in Part 1, this list includes glutamine, glycine, tyrosine, and taurine along with its precursor cysteine, derived from methionine. (154. Furst, 2000)

A couple more key amino acids that bear mentioning here are alanine (precursor to carnosine) & citrulline (precursor to arginine) which offer a range of benefits detailed in Part 6.

As illustrated by the amazing design of the human body, not only are many seemingly redundant backup pathways built into that design, but so too are many different functions of various proteins and enzymes along the enzymatic or biosynthetic pathways from precursors to end-product molecules.

For instance, glutamine is crucial for gut health and wound healing, whereas its mid-product α-KG regulates balance between anabolic and catabolic pathways, and its end-product glutamate is a vital neurotransmitter for the brain. (155. Deters, 2021)

Methionine and cysteine boost anti-inflammatory and neuroprotective benefits while their end-product taurine promotes proper bile synthesis; mitochondrial, heart, and brain health; blood sugar, fitness performance, immune support, and longevity.
(156. Jong CJ, 2021) (157. Walter, 2023)

Cysteine, glycine, and glutamine help form the antioxidant glutathione which is protective of most organs. Tyrosine promotes biosynthesis of dopamine, thyroid, and adrenal hormones, while its end-product, CoQ10 supports heart, kidney, liver, pancreas, lung, & brain health along with reducing inflammation and oxidative stress. (158. Childs, 2025) (159. Semeco, 2024)

The two flow chart diagrams below illustrate biosynthesis pathways for various metabolic processes. The first one shows all of the essential, conditionally-essential, and non-essential amino acids; along with a few other well-known antioxidants, hormones, and other metabolites. For simplicity and conciseness, many lesser-known metabolites and catabolites where one AA is broken down to form other AA pathways are not shown, but a few important ones *are* illustrated here with bi-directional arrows.

Glutamate and alanine, for example, are very central AAs responsible for multiple reactions in both directions. Dotted lines represent indirect enhancers; i.e., human growth hormone (HGH), which is not a direct metabolite of arginine, GABA, and lysine, but rather is stimulated or upregulated by these nutrients.

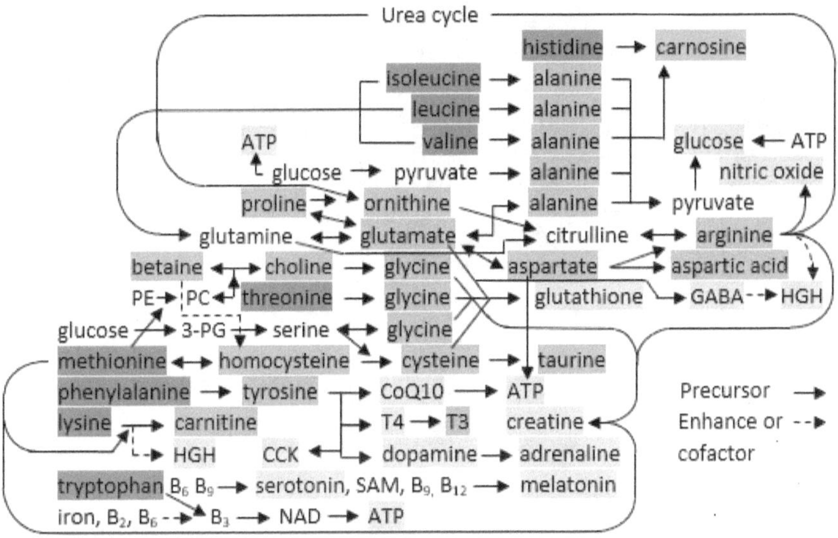

Figure 1. Amino Acids and Metabolites Biosynthesis

The nine essential AAs where pathways begin are highlighted in green or darker shade, whereas non-amino acid end-products or metabolites are highlighted in blue or lighter shade. AAs processed by the liver are highlighted in red or medium shade, illustrating the liver's central role in amino acid biosynthesis. If this seems a bit complex, just keep in mind that these schematics are just a comprehensive overview and several individual pathways will be broken out as simpler diagrams in later sections.

The second diagram below highlights some of the main pathways for hormone synthesis and energy production. As shown, certain minerals and B vitamins are crucial for energy production whereas cholesterol and P450 enzyme nutrients are essential and foundational for healthy hormone production.

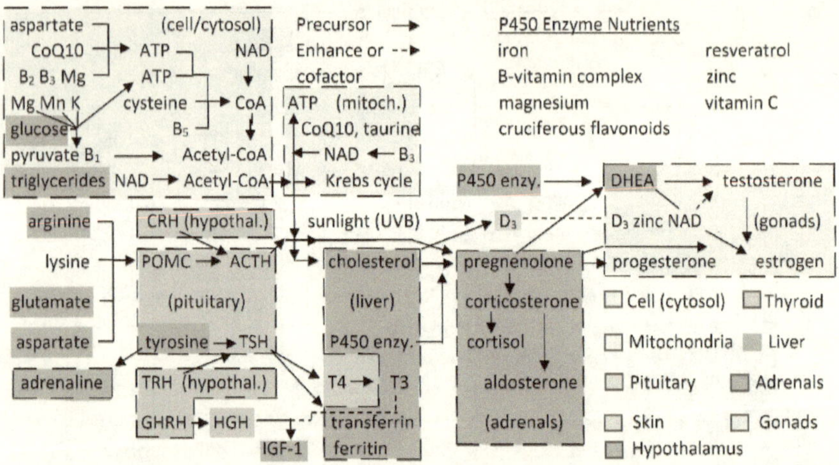

Figure 2. Major Hormones and Energy Production

As one can see, while supplements for certain amino acid end-products may be necessary or optimal for acute conditions or deficiencies, we should not underestimate the multiple benefits from biosynthesis of various nutrients derived from those lower down on the metabolite chain, such as essential amino acids.

However, healthy liver and kidneys are necessary for converting most essential amino acids to conditional and non-essential aminos, especially if conditionals and non-essentials are not being adequately consumed in the diet. That said, amino acids like glutamine, tyrosine, taurine, glycine, alanine and citrulline are all shown to exhibit great metabolic health benefits when supplemented by those who need it. Most amino acids can be found in meat, fish, poultry, dairy, legumes, nuts, and hearty vegetables.

Another interesting facet regarding protein foods is how essential aminos are often derived from not only animal products, but plant-based foods as well, while the further up the chain you go, the more likely aminos are to be obtained only from animal sources. The power behind animal protein is that they contain not only multiple amino acids, but also precursor and end-product aminos together.

Look, up in the cells, it's omegas....it's phospholipids!

As one of the most crucial building blocks for cells, chemical signaling, and mitochondria, we've seen how important phospholipids (PL) are for systemic health. PLs also support selective passage of contents in and out of cells for every cell in the body. Many PLs consumed for the body are ideally infused with polyunsaturated fats (PUFAS) such as omega-3 fatty acids, though some omega-6s are needed as well. In fact, PLs are the ideal carrier form of omega-3s. (160. Burri L, 2012)

Omega-3 infused PC can be obtained from wild fatty fish, pasture-raised eggs & liver, and algae, but farmed fish contain lower omega-3 content. In addition to brain health, omega-3s are shown to improve obesity, the liver, cellular senescence, and even arthritis.

At the apex of whole-body homeostasis, liver contains the highest profile of phosphatidylcholine (PC), aside from the lungs, followed by the heart, kidneys, and brain; whereas phosphatidylethanolamine (PE) comes in second for the liver, but first for the other vital organs except for the lungs where PG is second. (161. Christie, 2025)

Phosphatidylinositol (PI) and phosphatidylserine (PS) are two other PLs which carry similar significance for cell health, but are contained in the body in smaller amounts.

Monounsaturated fats (MUFAS) from sources like olive oil, avocados, and nuts are also critical when it comes to the brain, as are MCTs found in coconut oil, but these foods also contain phospholipids as well. Omega-3 infused PC found in fish, eggs, and liver along with MUFAS infused PE found in nuts and seeds are the most ideal PLs for the brain; eggs being especially high in PC. Non-MUFAS sources of PE include citrus juice and milk, but have low amounts unless consuming large quantities. (162. Züge, 2017) (163. Cui, 2015)

Taking this into consideration, consuming eggs with breakfast might be one of the best things you can do for your brain in the morning; along with eating fish, liver (if one can eat), nuts, and seeds throughout the day, especially for brain fatigued people like college students or those working long hours on the computer.

The gut benefits from PE because it makes up 90% of the phospholipid content in the G.I. tract. Along with eggs; soy, liver, beef, chicken, tuna and chickpeas are also high in both PC and PE; whereas eggs, salmon, and beef can be added to the list for foods high in PC, which is essential for the other organs such as the liver. PC can be biosynthesized from PE and methionine, but only 30% RDA of PC comes from this pathway. If deficiently warranted, PLs can also be obtained as supplements as well. (164. Wikipedia, n.d.)

Even though PG is minor in its overall composition of the body vs PC and PE, it is more abundant in the lungs, but it's real power may lie in its precursor ability to form cardiolipin (CL) which is vital for the very mitochondrial cell function itself including cell wall integrity, energy production, and a special mitochondrial autophagy called mitophagy. As a precursor to PG, PA is also a crucial factor for Cl and its best source is found in cabbage. (165. Urikura M, 2012)

Amazing new research also shows that PLs can actually help the body heal from mRNA toxicity, which may be welcome news to many suffering adverse consequences from the pandemic. According to one expert, PC and PE have the ability to encapsulate toxins within the cells and remove them from the body. Regardless of benefit, phospholipids will likely gain more traction in media in the near future. (166. Doniparthi, 2024)

Taurine and Choline – Brothers in Arms

Some nutrients are so impressive, they need to be expounded upon, and taurine is one of them. Adding choline to the mix bolsters this significance even more given the similar benefits each one provides. Although more like nutritional cousins than brothers, taurine and choline team up to support many systemic health benefits such as healthy digestion, metabolic homeostasis, muscle support, cellular integrity, and brain health. Taurine alone promotes other incredible benefits as well. (167. Johnson, 2019) (168. Thirupathi A, 2020) (169. Lee CW, 2023) (170. National Institutes of Health, 2022)

As one of the most abundant amino acids in the body, taurine is concentrated in the heart, brain, eyes, skeletal muscle, liver, and other body tissues; while choline acts more like a B vitamin or in other forms like PC that operates in these tissues as well. The liver is where they tag-team, but they also follow several parallel paths.

Recapping briefly from the Liver section in Part 2, taurine conjugates with incoming cholesterol bile acid substrates to form bile salts, after which PC jumps on board to neutralize pH and complete the bile process. Moreover, taurine suppresses oxidative stress and inflammation while choline and PC metabolize liver fat and mobilize triglycerides for transport to other parts of the body.

Although not actual building blocks like branch-chained amino acids, taurine and choline are absolutely essential for muscle synthesis and repair. Choline helps to maintain the integrity of the cell membrane, while taurine is crucial for maintaining the cellular membrane of the mitochondria within the cell, in addition to regulating electrical ion balance of the whole cell via potassium and magnesium inside the cell and sodium outside the cell wall.

Taurine also regulates calcium balance in cells and optimizes cellular energy by countering oxidative stress from muscle exhaustion. In fact, endurance exercise is shown to increase production and efficiency of taurine. As a muscle itself, the heart also benefits from taurine, especially under distress conditions such as in heart failure, where taurine increases energy via increased ATP production.

The brain is another area where taurine and choline work synergistically. Taurine buffers against inflammation, oxidative stress, and inhibits apoptosis (cell death), leading to neuroprotective benefits supporting cognitive functions like learning and memory. Choline likewise supports memory, cognition, and mood via the neurotransmitter acetylcholine.

In addition to all these functions, taurine also supports the pancreas by increasing glucose metabolism and improving insulin sensitivity, improves eye health, promotes longevity, and helps protect against metabolic syndrome, cancer, and CVD. Add to this increased muscle and endurance capacity and taurine truly lives up to its reputation as a super amino acid!

In summary, taurine and choline are a powerful combo with amazing benefits across the metabolic spectrum. Taurine can be found in high amounts starting with yellowfin tuna, scallops, mussels, whitefish, dark poultry, and red/green algae seaweed. Taurine can also be synthesized from cysteine-rich foods like pork, beef, chicken, fish, lentils, oatmeal, eggs, and dairy. Choline content via PC is high in beef liver and eggs, with moderate amounts in beef, soybeans, chicken, and fish, but can also synthesized from PE.

The Fantastic ~~Duo~~ Trio – Magnesium, B1, & B3

As some of the most popular nutrients among all ages, B vitamins and magnesium have certainly earned their wellness reputation. Other dynamic combos exist, but this one really seems to stand out.

Whereas B vitamins are hailed for everything from energy production, heart and nerve health, immune system & red blood cell support, skin, hair, and eye health, brain support, and just overall sense of well-being; magnesium is an amazing, hormone-like mineral responsible for 300 reactions in the body such as proper heart muscle and rhythm function, sleep and stress regulation, bone and muscle health, blood sugar and cholesterol management, energy metabolism, anti-inflammation, and calcification inhibitor.

For holistic support, magnesium pairs better with a complete Vitamin B spectrum, but as cofactors, B1 and B3 share a special synergistic relationship with magnesium, upregulating each other as opposed to competing with each other where an excess of one nutrient depletes another one. Vitamin D is also a cofactor with magnesium as it relates to calcium regulation, as previously mentioned in the electrolyte section of Part 2.

Vitamin B1 or thiamine is an extremely essential nutrient as shown by the historical scourge of beriberi caused from B1 deficiency, which prompted authorities to require processed bread makers since the early 1900s to add back in synthetic B vitamins to bread. The health benefits and importance of B1 can best be described by the many systemic-wide issues that can occur when B1 is deficient such as enlarged heart, edema, difficult breathing and sleep apnea, neuropathy, slow digestion, brain fog, insomnia, panic attack, fatigue, muscle loss and more. (171. Berg, 2018)

In short, B1 is critical for sustaining the autonomic nervous system which affects practically every system in the body. Adding to this criticality, B1 is only stored in small amounts from 2-3 weeks in the liver, so daily intake is needed and unfortunately, B1 is not found in as many common foods in high amounts than other B vitamins are.

That said, some of the best natural sources of B1 are rice bran, spirulina, seeds (flax, sunflower, hemp), nuts (pinion, macadamia, pistachio, pecan), pork, oats, salmon, peas, black beans, brown rice, and fresh-milled wheat, in that order by weight. Those on whole-food diets are encouraged to eat these or else supplement B1.
(172. My Food Data, n.d.)

Magnesium stands tall among all of the minerals because of its many intriguing functions and benefits, and in addition to those already listed, magnesium is also an important cofactor for sex hormones and fertility in both men and women. Similar to B1, the importance of magnesium is noticeable when deficiencies occur, but likewise, can be a challenge to get through diet if not consuming seeds, nuts, dark chocolate, leafy green vegetables, legumes, and coconut or mineralized water.

Magnesium also supports the hypothalamus-pituitary-adrenal (HPA) axis and helps to prevent adrenal fatigue. Fortunately, magnesium can be supplemented and tolerated in many chelated forms, each appearing to emphasize particular benefits. Listed below are five of the most popular and well-tolerated forms:

- Magnesium glycinate is very popular for its role in helping anxiety, insomnia, and sleep
- Magnesium citrate is generally used to help digestion and may have some laxative effects
- Magnesium taurate is known to help with high blood pressure and blood sugar regulation
- Magnesium L-threonate is used to support the brain with issues like depression and Alzheimer's
- Magnesium orotate is popular with athletes because of its role with energy and cardio support

  (173. Hill, 2023)

Magnesium and Vitamin B1 assist each other and share many overlapping roles, but the synergistic relationship is more than just observational. Research shows that magnesium is needed to help B1 convert glucose to energy inside cells, suggesting that magnesium is the limiting factor between the two, especially since magnesium can utilize another pathway for energy as well. (174. Overton, 2022)

This relationship is revealed by how both nutrients, among others, can be depleted or used up through the metabolization of sugar and carbohydrates, especially processed carbs, necessitating the daily requirements for both in the diet. In fact, these nutrients are often naturally occurring in foods like whole wheat and brown rice (which are processed to white flour and white rice) for that very purpose. This is one reason why consuming processed grains can be detrimental to our health over time. (175. Berg, 2025)

Another star of the show that cofactors with magnesium and other nutrients to produce energy is vitamin B3 or niacin. In fact, while B1 contributes to PNS functions and glucose energy conversion, B3 merges into the energy cycle in multiple places, not only in the cell at large, but also in what's known as the Kreb's cycle within the mitochondria itself.

As a precursor to the metabolite NAD, which has been popularized as a supplement on fitness ads, B3 is involved with mitochondrial energy production regardless of whether the enzymatic journey begins with either the glucose, amino acid, or triglyceride pathways. Like B1, magnesium teams up with B3 to enable this process as shown in the preceding Hormones and Energy Production chart.

Niacin's role in energy leads to many health benefits like increased muscle mass and decreased fat, enhanced muscle strength, and improved glucose metabolism, similar to taurine's effect.
(176. Xiang, 2023)

In addition to these capacities, B3 has been shown to exhibit many other support functions such as cholesterol regulation, brain health, skin protection against sun damage, and even alleviation of osteoarthritis and improved joint mobility. (177. Vinmec, 2025)

As three of the most essential nutrients present in all body tissues and necessary for not only daily energy, but a whole range of critical functions; magnesium, B1, and B3 are paramount for vibrant health and without a doubt, are an indispensable and synergistic super trio.

Vitamin C – The Unsung Hero of Brain and Tissue Health

Vitamin C is like the happy-go-lucky person that everyone knows, but is often taken for granted until someone is sick or in need. Many people remember taking vitamin C when they were young to ward off flu season, but then never giving it a second thought afterwards. Vitamin C is so much more than just an immune support.

In Part 2, we learned about some of the incredible benefits of vitamin C including proper glucose metabolism, relieving oxidative stress, improving brain cognition, and supporting blood vessels and myelin nerve sheaths. The antioxidant properties of vitamin C are crucial for vibrant brain health.

Zooming out to the whole body, we find that this vitamin also aids in wound healing and supports collagen production in connective tissue, muscles, skin, and even eyes, supporting eye health. In addition, vitamin C has been shown to lower blood pressure by assisting vasodilation of blood vessels and reducing stress by suppressing cortisol to relieve the adrenal glands, leading to positive outcomes like improved sleep and better weight management.

By contrast, adrenaline junkies and those wishing to weather the stresses of life will be happy to know that vitamin C is also a cofactor for converting dopamine to adrenaline. So, whether for boosting the immune system, enhanced tissue synthesis, or better adrenal function, vitamin C is a stealthy unsung vitamin hero we should all pay more attention to. (178. Geisinger, 2023) (179. Arora, 2025)

The Exotic Benefits of Zinc

As one of the most powerful and diverse trace minerals, zinc, along with its macro-mineral compadre magnesium, is considered one of the "sexy" minerals, in a literal sense. Responsible for a wide array of body functions, zinc is also crucial for supporting sexual function in both men and women, but especially men. (180. Brown, 2024)

Zinc, which can be found in shellfish, meat, nuts, legumes, grains, and dairy, is shown to improve ED, fertility, and testosterone levels which help to promote overall health and vitality. What's more, testosterone also helps to boost growth hormone (GH) synergistically since both hormones require zinc, and zinc is also required for foundational hormones like pregnenolone and DHEA, but especially testosterone. (181. Gibney, 2003)

More widely, zinc functions at the cellular level as part of the immune function, for DNA and protein synthesis, wound healing, vision and skin health, muscle growth, and other enzymatic reactions linked to digestion and reduction of fibrosis or scar tissue. Zinc supplementation has also been correlated with a reduction in atherosclerosis and CVD as well. (182. Kubala, 2022)

Two other minerals providing overlapping benefits are manganese and selenium which are often found together in a variety of foods similar to zinc. Manganese is noted for its support of cartilage and ligaments, while zinc, manganese, and selenium have all been targeted for cancer therapies. No doubt, zinc truly is an amazing and exotic superstar. (183. Rozenberg JM, 2022) (184. Yildiz A, 2019)

TUDCA – A Bear of a Supplement

At the forefront of organ homeostasis, the digestive system, which includes the liver and gallbladder, is crucial in helping to maintain vital health downstream. Tauroursodeoxycholic acid (TUDCA) is a remarkable bile acid supplement which is used primarily for digestive issues, but also other ailments as well.

Even the name itself reveals a clue about the powerful amino acid taurine which forms a substrate for this molecule. TUDCA, which was originally extracted from bear bile is now reproduced in labs and according to research is effective for healing the liver and reportedly, even dissolve gallstones. (185. Stong, 2022) (186. Song G, 2023)

In addition, TUDCA has also been shown to improve conditions in a wide range of diseases such as Alzheimer's, Parkinson's, and Huntington's, and eye disease along with stroke, diabetes, CVD, and kidney disease.

The main pathway in common for improvement in all these is the inhibition of apoptosis, an induced mechanism of programmed cell death for damaged cells as opposed to autophagy which attempts to repair the cells. UDCA is virtually identical to TUDCA, but administered as a medication, whereas CDCA and OCA, as noted earlier, are other bile acid options. (187. Vang S, 2014)

The many benefits and effectiveness of TUDCA underscore just how important naturally created bile in the body is. Science continues to explore new possibilities for treatments with TUDCA.

The Amazing Benefits of Berberine

One of the most fascinating supplements to gain popularity recently is a phytochemical found in several plants called berberine. Known for its healing properties in Asia for centuries, berberine has gained new traction in modern applications for ailments such as reducing NAFLD, lowering blood sugar, improving gut and heart health and cholesterol, supporting brain hormones for depression, suppressing cancer, reducing inflammation, and more. (188. Gunnars, 2023)

One of the ways it does this is by activating the enzyme AMPK which reduces oxidative stress and promotes autophagy or cellular repair in the mitochondria as opposed to apoptosis or cell death. Apoptosis is a necessary function to recycle critically damaged cells and prevent senescence, but can also lead to accelerated tissue damage if it proliferates without proper homeostatic regulation from certain proteins and enzymes. (189. Herzig S, 2017)

On the other hand, berberine can also support a different pathway by promoting apoptosis for cancer cells while protecting healthy cells according to research. While berberine is not for promoting muscle mass, its power lies in helping to suppress metabolic disease and inflammation through the catabolic pathways of autophagy and apoptosis, which is more in-line with longevity. (190. Zhu Y, 2022)

From anabolic-promoting pathways for strength and vitality to catabolic-promoting paths to prevent disease and inflammation, the body uses many nutrients to maintain optimal and vibrant health.

Part 4 – East Meets West:

From Ancient Holistic Insights to Modern Science

Since the beginning of societies on earth, people have strived to maintain the knowledge of food and medicinal plant therapies passed down to them by their Creator and ancestors, while also pursuing new discoveries through trial-and-error, observation, and investigating the physiology of the human body.

Many traditional therapies and knowledge of preparing, fermenting and balancing a variety of foods for maximum nutrition have been practiced right down to current times; even if a full spectrum of health benefits for some was not fully understood at the time.

The knowledge base may finally be coming full circle, and then some, as ancient divine insights and nutritional folk wisdom is being increasingly vindicated and corroborated through continuing advances in current scientific research. Many intuitive insights and methods from the past can actually be quantified and explained scientifically now, whereas once, they may have been relegated to the realm of conjecture and folksy tradition.

In Part 4, we will deviate from the current health context of our day and explore not only ancient views on health and nutrition, but also intriguing paths into cryptic history and spiritual related insights from a holistic perspective leading up to them. Stay tuned as we connect dots and consider what I believe are very credible concepts.

From every place on earth, people have adapted their diet and traditional therapies to the environment around them by eating a variety of foods and herbs available to them. Every culture has their own sage advice such as India with ayurvedic medicine or Native Americans with "Eat the organ you want to heal" wisdom, but for channeled insight that reaches back millennia, this research zeros in on a couple of ancient traditions or inspired perceptions from two regions – China, and the Levant or Eastern Mediterranean.

Many famous "blue zones" are from the latter region, including some not officially on this list, like the Caucasus. Except for a few dietary food parallels, the specifics of longevity blue zones will not be detailed here, but the internet is full of articles on this topic.

It should also be noted that some debate does exist in media over possible biases on this list (unintentional or otherwise) regarding the downplaying of meat consumption and/or accuracy of claimed ages of some very old centenarians. Nonetheless, there have been a few cases of extreme longevity recorded beyond 120 years.
(191. Robbins, n.d.)

That said, what all these locations — which have managed to evade the influx of ultra-processed diets — have in common are a diverse and healthy whole foods diet, lower stress, physical activity or outdoor lifestyle, and faith-based/social connectedness. In short, living more in harmony with nature and its natural rhythms the way we were created to appears to enhance better odds toward longevity.

Regarding diet, and with the possible exception of Okinawa where fermented veggies are consumed, these places share an important commonality besides varying degrees of meat, vegetables, and grains and that is dairy — especially fermented dairy. Beans and legumes are also a nutritional powerhouse consumed in all these cultures.

Another intriguing fact is that every one of these places, without exception, consumes the phospholipid PG from at least one of these known PG foods – soy, oats, spinach, pastured eggs, chickpeas, and barley; along with PG's precursor, PA found in cabbage, which is also eaten in these places. As discussed earlier, PG is a precursor to cardiolipin (CL) which is critical for mitochondrial function, ATP production, nerve signaling, mitophagy, and disease prevention.

Far from being anecdotal, these longevity correlations appear very compelling, in addition to the antioxidants and other nutrients these diets contain. While dairy consumption has always been a staple in the mediterranean, middle-east, and Caucasus; history shows that certain segments of ancient Chinese society included dairy as well, though Chinese culture is more widely recognized for its holistic system of healing, herbs, and physiology called traditional Chinese medicine or TCM. (192. Fuentes JM, 2024)

Commonly known for Chinese herbs and acupuncture, TCM has also identified links between emotions and organs with respect to ailments and disease since antiquity. Even more intriguing is textual evidence detailing similar connections in the ancient Levant as well.

To some, these ensuing ancient conclusions may not seem precise by today's standards, but are proven to be remarkably insightful given the lack of modern technology available during those eras. The following sections will explore these fascinating perceptions as it relates to western medical discoveries today, but first, an intriguing overview of ancient historical backgrounds, traditions, and evidence will pave the way for context leading into each topic.

Traditional Chinese Medicine

In the Beginning – The Chinese civilization is one of the oldest continuing cultures on earth with amazing insights regarding the human body, energy work, and herbal medicine; the foundations of which, may have been conveyed directly from God through the post-flood patriarchs to leave a lasting imprint deep in the consciousness of Chinese culture from antiquity.

To bolster these assertions, let's take a mini-dive into anthropology and cryptic history not commonly discussed in academic circles. Though sparce, there are historical inferences linking the Chinese all the way back to Mesopotamia and ancient biblical times.

On one hand, the Table of Nations describe genealogical records identifying the origins for what are now, many well documented ancient Mesopotamian cultures. On the other hand, ancient Chinese records, though often shrouded in what appear to be obscure mythological legends, go all the way back to around the same time period in the 3rd millennium B.C.

In the Bible, the Table of Nations describes many known tribal lineages, but that of the Chinese has always been harder to pin down. The best candidate put forward by many throughout history is Sin or Sinim, father of the Sinites. Though not cited by name in Genesis, Sinim is revealed through variations of the root name Sin both in the Bible and other sources including places like Sinai, Sian or Xi'an, Qin, and the term "sino" as in Japanese-Sino relations. (some render Sinim as Syene, referring to Aswan — itself still carrying the Sin root)

Sinim *is* mentioned however, in Isaiah 49:12 which states "Behold, these shall come from far: and, lo, these from the north and from the west; and these from the land of Sinim." While some documents relate them to Siannu on the Levant coast or the Desert of Sindh and Sindhu River in India, this may simply explain how sub-groups of Sinim broke off from a main branch which migrated south from Babylon toward Egypt after the confusion of languages and dispersion of peoples listed in Genesis 11. (193. Creationwiki, 2025)

While Sino Babylonianism, a theory of mass migration directly into China from Babylon has met with skepticism, the more likely scenario is that the Sinites left and passed through places like the Wilderness of Sin and Mt. Sinai (where Israelites later sojourned) before or after migrating to Egypt, on their way along what would later become known as the Maritime Silk Road; perhaps hugging the coast along the Indian Ocean before finally arriving in China.

Geoscientist Sun Weidong inadvertently made this connection while discovering how some very ancient Chinese bronze implements are chemically identical to Egyptian bronze as opposed to native Chinese bronze. Moreover, it's not clear whether the Sinites were pushed out or left voluntarily, but an ancient historical text referring to an ancestral home containing a north-flowing river which divides into nine tributaries before flowing into the sea, perfectly describes the Nile River Delta. (194. Lewis, n.d.)

Lineages, Myths and Legends – In Genesis 10:17, the Sinites are listed as a tribe from one of eleven sons of Canaan, son of Ham, son of Noah. Conceivably, the bulk of Sinites may have sought to distance themselves from their more pagan Canaanite brothers — who were squatting on Shem's allotted portion — as shown by their ancient monotheistic roots and eventual migration east into China.

Obscure legendary clues may point to their veneration of Sin's great-uncle Shem and his father Noah over direct ancestors Ham and Canaan. The ancient collection of rulers called The Three Sovereigns and Five Emperors describes an intriguing account of three extremely long-lived god-like rulers who taught people the basics of life and laid the foundation for Chinese culture, followed by five emperors who were model rulers for advancing society.

Fuxi is the heavenly sovereign who taught people the tenants of marriage and basics of subsistence. Though mythologically obscure, Nuwa (sometimes a man) is the earthly sovereign and Fuxi's wife.

After a great flood, Nuwa was said to fix the sky with seven colored stones (rainbow?) and is responsible for reproducing people. Shennong or Shen Nun is the "divine farmer" who taught people how to farm and identify plants and herbs.

The Yellow Emperor Huangdi, first of five emperors and ancestor of the Han Chinese, is also identified with the three sovereigns as a relative of Shen Nun. He is credited with inventing copper-ware, a compass, boats, calendar, and writing. Emperor Zhuanxu was the grandson of Huangdi who led an eastward migration and enlarged his tribe. Emperors Ku and Yao were descendants of Huangdi, but Emperor Shun, who was chosen by Yao to succeed him, was not. Shun is the first reported to have offered the Border Sacrifice to Shangdi, the God of Heaven. (195. New World Encyclopedia, n.d.)

In other accounts, Fuxi, Shen Nun, and Huangdi are identified as the three sovereigns implying that Huangdi is a transitional figure between the "Three Immortals" and other four emperors. Unlike the semi-gods Fuxi and Shen Nun, Huangdi's human birth may likely be a mythical reference to Sin's birth on post-flood earth as opposed to Noah and Shem who stepped off the ark from a previous world. (196. China Sage, n.d.) (197. Vasilev, 2021)

In light of this connection, it's also likely that Fuxi is actually Noah whereas the name Nuwa was misappropriated to Noah's wife. If this all sounds like a stretch, stick with me, another culture related to the Han Chinese reveals fascinating corroboration with this theory.

Unlike the Han, the Miao people — of which a portion of them merged with the Han Chinese in ancient times — report a concise history of their ancestors which is not cryptic at all. Much like the Bible, they recount the creation story, the great flood; and how Nuah with his three sons Lo-Han, Lo-Shen, and Jah-hu survived the flood; along with the account of the Tower of Babel. It's clear that Nuwa or Nuah refers to Noah and Lo-Han, Lo-Shen, and Ja-hu to his three sons Ham, Shem, and Japeth, respectively.
(198. Truax, 1991) (199 Asia Harvest, n.d.)

Furthermore, it's conceivable that Shennong or Shen Nun is the same person as Lo-Shen or Shem. Also, the historical capital of Shaanxi province, Xi'an, was formerly called Sianfu which is translated to "Father Sin". All a coincidence? I find it compelling.

As the father of the Han Chinese, is it possible, even probable that the Yellow Emperor is Father Sin himself who brought, or whose descendent brought Egyptian bronze to China in boats using a compass? It's a fascinating topic to explore, but whatever the case, it's plausible that the ancient Chinese did revere Noah, Shem, and Sin; but also worshipped the God of Heaven through annual border sacrifices to Shangdi, the "Heavenly Ruler".

To summarize, with all this inferred evidence, I believe it's credible to assume that the basis of early Chinese culture, and by extension, at least some elements from the earliest roots of TCM are grounded in Middle Eastern origins and divine inspiration from God.

The Advent of TCM – Building on Shennong's legacy of testing and cataloguing of many plants and herbs, the Yellow Emperor (Huangdi) is generally considered the founder of traditional Chinese medicine who is believed to have invented acupuncture and received insight about the pulse and other diagnostic tools from "immortals", who we now see may be construed as the patriarchs.

Credited to Huangdi, but logged much later, actual documentation and compilation of TCM as a more pragmatic practice is said to have begun around the 3rd century B.C., but greater awareness and elevation of its status appear to have begun earlier around the 5th century B.C. as reflected by great physicians like Bian Qiao and others. The concepts of Qi, yin and yang, five symbolic elements (wood, fire, earth, metal, and water), theory of organs, acupuncture meridians, and standard use of herbs seem to be codified during this period according to sources. (200. Britannica, 2025)

It is interesting to note, that in biblical terms, the number five is associated with grace, favor, and natural order of creation, so it's intriguing to see how this number is revealed through certain aspects like the five elements, five senses, five fingers/toes, and even the bible. The Ten Commandments are seen as two sets of five; five dealing with a vertical relationship to God and five with horizontal relations to other people, whereas the 50 year jubilee is a multiple of five (and ten). (201. Fagan, 2024)

Though diverse schools of thought and beliefs integrated with TCM over time, the basic concepts outlined above which describe the oscillation, regeneration, and balance in nature; along with ties to human health within the created order, originate from early contributors to TCM and are summarized in the following sections.

Qi (Vital Energy) – The overall concept of Qi includes the energy force all around us, which according to modern string theory in quantum physics, includes all matter as well. In TCM, Qi not only binds together the universe (Colossians 1:17), but also facilitates communication between nature and its surrounding environment as seen with the seasons, tides, landscape, and biological life. Likewise, Qi is also seen as the governing force that works through meridians of energy linking together all the functions of the human body. (202. TCM World Foundation, n.d.)

Yin and Yang – The concept of yin and yang integrates opposing, but complimentary forces in nature that define the ebb and flow of life to help maintain harmonious balance. Examples are seen in the contrast between summer and winter, day and night, male and female, hot and cold, and active vs passive as illustrated by our sympathetic and parasympathetic nervous systems along with our circadian rhythms. Yin and yang are also seen in opposing features like elements, directions, and shapes; i.e., fire vs water, north vs south, and round vs flat. Together, Qi, along with yin and yang, lead to the next progression in natural cycles called the "five element theory". (203. TCM World Foundation, n.d.)

Five Elements – In TCM, the five elements describe the dynamic relationships between five core elements in nature which include wood, fire, earth, metal, and water in repeating cycles known as the generative or productive cycle, and supportive or controlling cycle.

For instance, in the controlling cycle, wood (plants) stabilizes earth, earth channels or contains water, water controls fire, fire forges metal, and metal controls or shapes wood; whereas in the productive cycle, wood feeds fire, fire creates ash to nourish the earth, earth produces metal, metal generates water, and water nourishes wood. Both cycles are shown below. (204. The FengShui Academy, n.d.)

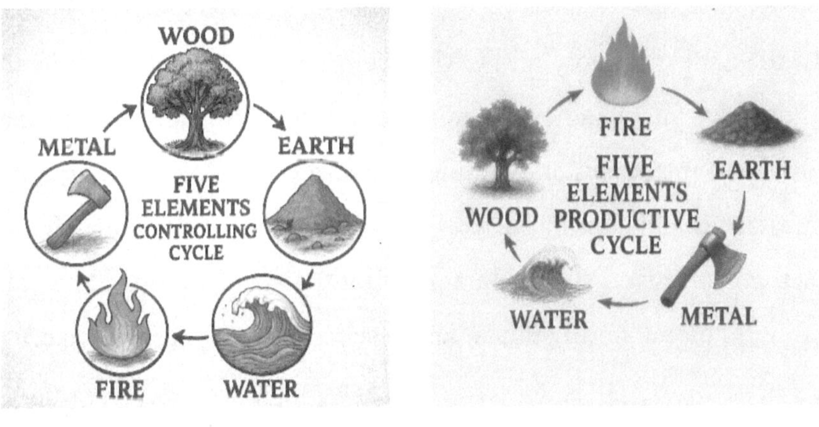

Figure 3. Five Elements Controlling and Productive Cycles

At this point you may be asking, "how does metal lead to water?" Well, interestingly enough, atmospheric science confirms that a large portion of rain dust particles called cloud condensing nuclei contain metal ions such as magnesium, aluminum, sodium, and even iron which help to precipitate rain drops. (205. Wikipedia, n.d.)

Was the connection between metal and water just an imaginary stretch to complete the five elements cycle or did early Chinese patriarchs actually have higher knowledge of this phenomenon? It's hard to say, but if so, this may add credibility to other holistic developments such as the theory of organs and others.

Theory of Organs – This study of anatomy incorporates the concepts of Qi, yin and yang, and the five elements to help define the complex interrelationships between the body's organs. The theory of organs is a foundational theory used for diagnostic troubleshooting and treatment modes such as pulse-taking, acupuncture, herbal medicine, and other remedies.

In the zang-fu theory of TCM, certain organs are considered yang and others yin. Yang organs such as the stomach, gallbladder, small and large intestines, and bladder are responsible for breaking down substances and eliminating waste products; whereas yin organs such as the heart, liver, spleen, lung, and kidney produce, transform, and store qi, blood, bodily fluids, and essence. (206. Schoenbart, n.d.)

Modern TCM researchers have back-tested and consolidated zang-fu down to five yin and one yang organs representing the liver, heart, spleen, lungs, kidneys, and stomach, respectively; using correlation analysis of disease syndromes based on recorded data from historical dynasties. Accordingly, the gallbladder is associated with the liver and stomach, the small and large intestines with the stomach and spleen, and the bladder with the kidneys. (207. Zhang L, 2011)

This makes sense since the G.I. tract is treated as a whole, bile from the liver and gallbladder do affect the stomach and duodenum, and of course the kidneys both empty into and affect the bladder. The correlation between the stomach and spleen is less obvious except that nutrient-enriched blood from the G.I. tract actually merges with pathogen filtered blood from the spleen and return blood supply from the pancreas and gallbladder. (208. Medline Plus, 2024)

This blood supply is shuttled through the portal vein which feeds into the liver, so in that sense, there is a connection. Moreover, since 70-80% of the body's immune system resides in the gut, this links the G.I. tract together with the lymphatic function of the spleen, as co-immunity organs. Combining the stomach and spleen as one element then reduces the six zang-fu organs down to five elements.

Though not originally addressed in TCM, the pancreas would likely be associated with the gut and/or liver and the thymus perhaps with the spleen. The adrenals are associated with the kidneys in TCM, which makes sense because of hormone and electrolyte connections, but are not specifically referenced within the five elements.

In nature, the five elements are associated with the five main organs (and emotions) represented by both the controlling cycle and productive cycle. The order in the productive cycle is: wood/liver, fire/heart, earth/stomach(spleen), metal/lungs, and water/kidneys. The metabolic correlation in this order appears weak, but where it really starts to get interesting is with the controlling cycle.

In the controlling cycle, the organ/element associations remain the same but the order changes to reflect: wood/liver controls the stomach, earth/stomach controls the kidneys, water/kidneys control the heart, fire/heart controls the lungs, and metal/lungs control wood/liver as shown below. The associated elements are wood controls earth, earth controls water, water controls fire, fire controls metal, and metal controls wood. (209. TCM World Foundation, n.d.)

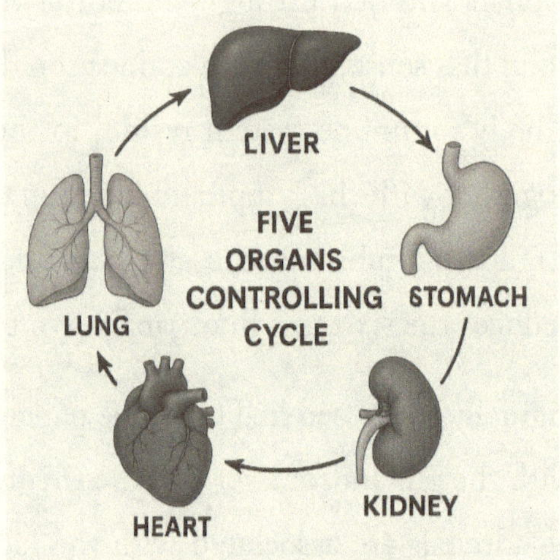

Figure 4. Five Organs Controlling Cycle

Indeed, from all the research presented and the chart above, we can see that the liver and gallbladder do impact the gut, thyroid, and adrenals which in turn affect kidney regulation; impaired kidneys contribute to atherosclerosis and poor heart health; the lungs depend on nutrient-enriched blood flow from the heart; and the liver depends on oxygenated blood supply from the lungs. But why the focus on just oxygenated liver? Another convenient placeholder?

Well, as it turns out, while every organ depends on blood and oxygen, the TCM focus on lungs and liver in this regard is actually backed up by science. As the only oxygenating organ, the lungs receive the entire de-oxygenated blood supply from the heart, and it just so happens that the liver receives the highest proportion of oxygen at 20-24%, more than any other organ including the brain! In light of current scientific research, the correlations with TCM organ theory are really quite amazing. (210. Jrank, n.d.) (211. Jahn, n.d.)

Despite cultural taboos limiting access to cadavers in ancient China, which prevented physicians from performing autopsies for better anatomical clarity, TCM accuracy with regards to how organs interact is pretty remarkable. In addition to anatomical insights, each organ is also associated with a particular emotion, along with other attributes within the construct of five elements.

The human body represents the most complex control system design in the universe, and while metabolic effects of organs tend to follow the controlling cycle, emotions tend to follow the productive cycle, with fire or joy leading as the starting reference point.

The heart is connected with joy and the element of fire. Respectively, the stomach and spleen are negatively linked to worry or anxiety, but also positively as "singing" and the element of earth. The lungs are related to grief and the element of metal. The kidneys are associated with fear and the element of water. Lastly, the liver is associated with anger and the element of wood.

While day-to-day emotions may randomly follow any order, the cyclical nature of emotions in TCM through events and seasons of life goes something like this:

The release of anger gives way to joy and empowerment. As joy slowly diminishes into contentment, and then discontentment, doubt and worry creep in. Worry is finally confirmed by the realization of circumstances that leads to grief. Grief gives rise to fear of future loss or uncertainty. Fear is then replaced by protective anger and a resolve to move forward out of vulnerability.
(212. Felicia, n.d.)

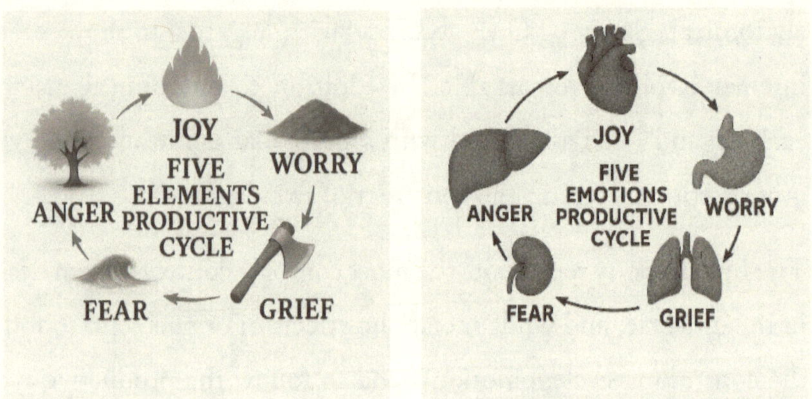

Figure 5. Productive Cycle Associations Between Emotions and Organs

The organs in this pattern are not metabolically connected in order, but rather are simply associated with and affected by the emotions as they move through each stage of the cycle. For more traumatic or perilous events, grief may be short-circuited by fear and placed after anger as a delayed response in the cycle.

This concept does not appear to be specifically addressed in TCM, but is commonly recognized today as the five stages of grief: denial (fear), anger, bargaining, depression (grief), and acceptance (contentment, akin to joy). If bargaining, an intermediate between anger and grief, is removed and an initial event emotion like worry inserted before denial (fear), this may represent an alternate version of the five element emotion cycle. (213. Bethany, 2024)

Coming full-circle back to the end of Part 1, we see particularly how issues like childhood trauma, PTSD, and repressed anger can degrade liver function, as supported in theory by TCM which associates anger with the liver and gallbladder. We also see how worry can affect the stomach and grief can affect the immune system (lungs), along with the other correlations laid out above.

In summary, the concepts of five elements and organ theory illustrate the intriguing cyclical patterns not only in nature at large, but also within the human body and psyche, thus leading to the development of a systematic baseline for understanding and treating ailments through traditional Chinese medicine techniques.

Acupuncture Meridians, Pulse Taking, and Herbs – The very progression of TCM theories that identify organs with emotions, nutrition, and disease sets the stage for acupuncture meridians, pulse taking, and herbal medicine for diagnosis and treatment. A deep dive into these topics is reserved for clinical experts and beyond the scope of this book, but are summarized here.

Essentially though, meridians are described as pathways of energy, which in modern terms, likely correlates with the nervous system, but perhaps also other cell-to-cell communication pathways that are yet to be fully understood by western medicine. Meridian pathway diagrams are highly detailed and, in many ways, resemble modern anatomical charts that one would see in a doctor's office.

There are twelve major meridians on each side of the body that mirror each other. Each organ not only corresponds with a specific meridian, but is also connected with the other organ systems and meridian network as a whole. (214. TCM World Foundation, n.d.)

Before treatment, a practitioner may assess the texture of the tongue and evaluate not only the rate of the pulse, but also the strength and tone as well. The pulse and tongue are two indicators that convey the current state of health and balance within the system. Symptoms are resolved by treating root causes of ailments. Western doctors used to evaluate the tongue, but that practice seems to have waned.

Personal experience affirms that acupuncture works and can be used to help calm the body, relieve pain, and unblock or re-energize "stuck" organ systems that have become metabolically unhealthy.

Of course, this is just one mode along with proper nutrition, exercise, and lifestyle changes to advance toward more optimal health, but it can be a good tool to use in parallel for helping to speed or facilitate the healing process, especially early on when conditions are more acute and severe. (215. Bhanote, 2023)

One final mode of treatment involves the prescription of herbs which have been classified and used for thousands of years. As cited earlier, by either oral tradition or manuscripts lost to time, Shennong or Shen Nun is credited for testing and cataloguing many plants and herbs, but the field of study was greatly expanded by many ancient leaders and practitioners who came after him.

Chinese herbal medicine (CHM) addresses practically every part of the body and ongoing research is continually being conducted trying to verify and synergize eastern medicine with current western medicine practices. While many herbs are familiar only to practitioners, many others have become popularized in mainstream culture for their medicinal benefits. Some can be consumed on a regular basis whereas others need to be cycled.

Cordyceps are medicinal mushrooms used to treat kidney disorders, improve lung infections, support reproductive health, stop bleeding, and increase energy. Ginseng is a widely known herb said to help with diabetes, fatigue, palpitations, shortness of breath, insomnia, impotence, digestion, and support all of the five-element organs. Reishi is an energizing mushroom known to promote stamina, vitality, protect the liver and heart, and even slow aging.

The benefits of goji berries include supporting eye health, blood sugar levels, mood balance, fertility, liver detox, and fighting cancer. Schisandra has been used by TCM practitioners to work through meridians and help balance energy in the body. (216. Axe, 2023)

Ancestral Diet and Dairy (or lack of) – Historically, the Chinese diet has consisted of a variety of foods, following along the timeline with rice, millet, wheat, soybeans, bok choy, broccoli and other vegetables in that order; along with pork and also beef and mutton which followed soon after. Only wheat and millet could be grown in the colder and drier northern part of China, whereas rice was cultivated in southern China which is warmer and wetter. Meat was scarcer for common people and more reserved for the wealthy.

Except for earlier influences from the north and in more recent times, dairy has been largely absent from the traditional Chinese diet. However, historical records reveal that dairy consumption was more prevalent among the aristocracy and Buddhist monks during medieval times, and from Mongolian and Khitan or Cathay influences north of China, but fell out of favor for centuries until recent decades. (217. Adhikari, 2024) (218. Kotyk, 2021)

Chinese culture has always been acutely aware of its historical past influencing society for good, bad, or sub-optimal and dairy is no exception. Beginning with European colonization in the 19th century and culminating in the 1980s with new market reforms and foreign exposure via 1984 Olympics, Chinese society began taking note of how milk and dairy contributed to foreigners being taller and stronger. This sentiment was undeniably expressed by the Chinese government's ensuing efforts in promoting a vastly accelerated increase in its dairy industry.

While visiting a large dairy factory in 2014, the Chinese President reportedly expressed that the "new China man" is expected to be a milk-drinker, reflecting the national view of dairy playing a strategic role in building a stronger and healthier nation. (219. Lawrence, n.d.)

The story of Chinese culture and its impacts on the world is a long and ancient one. From deep ancestral insights grounded in monotheistic inspiration, to TCM, to modern recognition of dairy as a strategic national priority, Chinese society and traditions have enriched the world's understanding of health and well-being through a wealth of knowledge and insight from ancient eras right up till current times.

As one of the oldest continuing civilizations today, China's ancient written records may not be as old as other ancient writings from Mesopotamia and the Levant; but its legendary accounts and keen perceptions on human physiology and the surrounding environment reveal clues of an original understanding for living in balance with nature in God's created order.

The Levant, Patriarchs, and Ancient Israel

Historical Introduction – Mesopotamia is widely accepted as the cradle of civilization. The oldest recorded writings and remains of civilization can be found in this region, which in ancient times, encompassed the entire area surrounding the Tigris and Euphrates rivers in parts of ancient Iraq and eastern Syria.

Ebla tablets found in northern Syria also reveal slightly less ancient writings and culture in this area of the eastern mediterranean or Levant bordering northwestern Mesopotamia. The overlap of these two regions contains Haran, one of the oldest cities of antiquity near the ancestral home of Abraham. Abraham's line descends directly from Shem – Ham's brother, Canaan's uncle and Sin's great-uncle.

While the ancient Levant is not typically referenced with regards to health or physiology, a wealth of knowledge and insight can be gleaned from ancient biblical manuscripts like the Bible. Among them are also Dead Sea Scroll manuscripts little-known to the general public called "The Testaments of the Twelve Patriarchs".

In them, Jacob's twelve sons record their autobiographies and admonitions to their sons for right living before they pass on, along with some prophetic warnings as well. Moreover, and similar to TCM physiology, three of the sons convey some striking insights regarding emotions and the effect on certain organs roughly 1200 years before the advent of established TCM.

Whereas the generally accepted timeline for TCM begins between the 5th and 3rd century BC, the era for The Testaments dates back to around the 18th century B.C., although the last originally recorded Dead Sea Scroll manuscripts in ancient times are dated to about the 2nd century B.C. (220. BiblicalTraining.org,, n.d.)

In general, some pseudepigraphal sources, which may or may not be faithful representations of original degraded texts or oral traditions, need to be taken with a grain of salt and evaluated against scripture and authentic writing style at that time; however, when they triangulate with other texts, allusions in scripture, and Dead Sea manuscripts, some robust conclusions can be gleaned from them.

The current version of the book of Jasher for instance, is thought to be an embellished revision of the original lost manuscript, whereas the Testaments of the Twelve Patriarchs seem to align much more with the aforementioned criteria and also happen to be apocryphal in the original Armenian bible.

Moreover, it's reasonable to assume that the original Essenes at Qumran (as opposed to A.D. gnostic Essenes) who recorded accurate messianic prophecies a couple centuries before Christ, also applied due diligence in striving to faithfully ascribe the accounts of "The Testaments", although some editorial conclusions may have been added as well. (221. Johnson, 2017)

Armenia, the first official Christian nation, incorporated these earlier Dead Sea scroll texts into their bible between the 4th and 5th century A.D. and in recent times, new batches or copies of nearly identical texts have been discovered which corroborate the originals that were thought to be less credible. In short, The Testaments do appear to satisfy the rigors of scrutiny, especially with certain details and dates that triangulate with other texts. (222. Andrews, n.d.)

Of course, the bible itself which is much more expansive and spans a much greater period of time contains many other clues regarding health and nutrition, so for more insightful context, we'll now proceed with a summarizing overview of these historical settings.

Geographical Context – Abraham, known as the "father of many nations" including Israel (Genesis 17:5), is the first Hebrew patriarch to enter what was then called the land of Canaan by crossing over the Euphrates River, from which the term "hebrew" originates (to cross over). Blessed by God, Abraham fathered not only Isaac who fathered Jacob, the father of the twelve sons of Israel, but also the Ishmaelite and Arab nations.

Collectively, Abraham, Isaac, Jacob, and his twelve sons are known as the biblical patriarchs who received God's Covenant and lived in the land from approximately 1875–1660 BC before Jacob and his family moved from Canaan to Egypt during the great famine. This correlates with 430 years from the call of Abram in Ur/Haran (possibly 1880 BC) to the time of the exodus which syncs up dates with both the Bible and the Testaments of the Patriarchs.

These dates follow the Septuagint chronology and Testaments as opposed to the Masoretic text, and mathematically speaking, are validated by a more realistic timeframe in Egypt and a much higher exponential decay correlation for decreased life spans from Noah to Moses ($R^2 = 0.97$ vs. 0.87), but that's a subject for another book.

Whereas current traditional thought favors a departure from Ur in Sumer, increasing evidence points to Ur in Anatolia (modern Urfa) as his ancestral home since Haran is only 25 miles away (versus 600) and Abraham's brother was named after Haran. Furthermore, Abraham's grandson Jacob is later called a "wandering Aramean" in Deuteronomy, possibly referring to either his time spent in Haran or maternal lineage. (223. Rendsburg, 2021)

From a geographical perspective, the Patriarchs' sphere of influence, including cultural and dietary similarities, encompassed the ancient Levant or what we now call Israel, Lebanon, southern Turkey, Syria, Jordan, and northwestern Iraq with cultural affinities possibly extending to the Caucasus via ancient Hurrians and other groups.

The Patriarchs were largely nomadic and their diet seems to overlap with foods found in both the Mediterranean and Paleo diets. In what could be termed the Paleo-terranean diet, it first and foremost consisted of grains, then dairy, meat, and legumes along with some vegetables, fruit, nuts, and occasional fish, as will be shown later.

Depending on portions and availability, it may have been a fairly well-balanced diet, and owing to their feats of physical strength and longevity; the patriarchs most certainly were not mal-nourished. In fact, diet may have been a major contributor to their prowess besides genetics. Better context is given to their physical environment by zooming in on the Patriarchs migratory movements, geo-political challenges, and battle engagements.

Heroic Faith and Epic Sagas – Genesis 11:31 states that Terah, along with his son Abram (later Abraham) and the rest of his family set out for Canaan, but stopped and settled in Haran along the way. Whether Terah was influenced by Abram in this decision is not specified, but we know they dwelt there for some time until after Terah passed and Abraham resumed course for Canaan at age 75.

At the crossroads of major trade routes where raids were an issue, Haran could be where Abraham acquired and/or added to his household and 318 trained men before setting out for the land of Canaan, where God would bless him and make him into a great nation (Gen. 12:1-3). The extra-biblical book of Jubilees (10:22-26) may offer some insight on this regarding dispersion and bestowed lands after the Great Flood. (224. The Firmament, n.d.)

While Jubilees was written or copied in the 2nd century B.C., and is canonical only in the Ethiopian bible, the section describing regional allotments of the earth to Noah's sons and grandsons does appear to have merit. The book of genesis describes the scattering of nations to various lands, and as a recipient of Noah's curse, Canaan does appear to be squatting on Shem's land as shown by God's eventual expulsion of the Canaanites from the land by Israelites for their wicked behavior centuries later.

By faith, Abram moved his entire household to the troubled land of Canaan promised to him by God, and throughout Genesis, he surveys the land by relocating in nomadic fashion to various places.

Sometime after living in the land, Abram takes his 318 trained fighting men to go rescue his nephew lot from kings of the east who had taken him captive along with spoils from battle with Sodom, Gomorrah, and other local kings. Using night warfare, Abram pursues, attacks, and routes a much larger army coalition in order to recover Lot and others with him. (Gen. 14).

The mysterious king of Salem, Melchizedek, blesses Abram for his righteous courage and Abram reciprocates by tithing a tenth of his spoils, recognizing God's favor in his victory. Later, renamed by God in a covenant, Abraham begets his son Isaac at the age of 100. The inheritance passes down to Isaac and then to Jacob, whose two sister-wives Leah and Rachel, along with their handmaidens bear him twelves sons and one daughter during his time spent in Harran.

Upon returning home, Jacob settles his family and household near the town of Shechem, where the ruler's son violates Jacob's daughter Dinah, sealing the fate of that town's destruction and a series of clashes with neighboring Canaanites following the incident.

Infuriated by the violation of their sister, two of Jacob's sons, Simeon and Levi, kill all of the male inhabitants of Shechem while they are recovering from an agreed upon rite of circumcision proposed by the sons in exchange for intermarriage as a path for peaceful coexistence. An exacerbated Jacob exclaims to his sons that he has now become loathsome to all the surrounding peoples when he realizes there will be constant hostility going forward (Gen. 34).

The bible is silent about any ensuing campaigns in the area and only mentions that a fear of Jacob's household from God fell over the surrounding towns when they finally did leave the area. The Testaments and other texts however, describe epic battles where the outnumbered sons of Jacob destroy wave after wave of offensive attacks from local coalitions. These texts reveal some fascinating insights into the lives and challenges of Jacob's twelve sons.

The Testaments of the Twelve Patriarchs – As briefed in the historical introduction, the Testaments describe each of Jacob's son's life accounts, exploits, confessions, admonitions for righteous living, and even prophecies to their sons before they pass on. By this time, they had long resettled in the land of Goshen in Egypt and the whole saga where they sold their brother Joseph into slavery who eventually became leader of the land, had long passed.

The accounts are recorded by order of the sons' birth and while not every son will be covered here, the ones that highlight key points regarding health and vitality will be. For completeness, the twelve sons listed in order are: Rueben, Simeon, Levi, Judah, Issachar, Zebulon, Dan, Naphtali, Gad, Asher, Joseph, and Benjamin.
(225. Knight, 2023) (226. The Firmament, n.d.)

The firstborn son Rueben recounts his testament and life events while admonishing his sons on how the "spirit of fighting in the liver and the gall" can result from a "spirit of error" if anger is not properly guarded.

Likewise, Simeon recalls being sad and repentant while being "pained in the liver" as a delayed consequence of jealousy and hatred he felt for his brother Joseph before they sold him into slavery.

Judah agreed to sell Joseph to prevent him from being killed by Simeon and Gad, but doesn't appear to have suffered illness by storing wrath like some of the others. As the preeminent son, though not first-born, Judah (fourth son) shares some amazing personal feats during the family's sojourn in the land of Canaan.

Judah reminisces about how he was able to overtake wild game, slay wild predators by hand, and lead his brothers to return hostility by waging epic battles against fortified towns though vastly outnumbered. Thus, his tribal emblem as "the lion of Judah" foreshadowing his messianic descendent almost two millennia later is confirmed by Jacob's blessing (Gen. 49:9).

Issachar is noted as being a farmer like his grandfather Isaac in contrast to his brothers who pastured flocks. Details are scant, but he testifies to raising seasonal crops for which his father was grateful, that likely included grains, fruit, and some vegetables.

Zebulon is more compassionate toward Joseph and testifies to preventing his brothers from killing Joseph. He also claims to be the first one to sail a boat in the region as he catches fish to feed his family and those in need for a period of five years during a famine; logging the first record of dietary fish in the Levant. Zebulon's connection with ships is confirmed by Jacob's blessing (Gen. 49:13).

Gad recounts taking on wild beasts such as lions, wolves, leopards, and bears while guarding flocks. Like Simeon, he also confesses the consequences of his hatred for Joseph as he recalls almost dying from an organ disease that lasted eleven months. Some sources mention the heart while other mention the liver.

Naphtali was considered gracious and spoke "good words". His account is also the most revealing regarding physiology. His father appointed him as a messenger because "he was swift as a deer" and he is associated with that animal in Jacob's blessing (Gen. 49:21).

In his testament, Naphtali recognizes the five senses and associates the heart with understanding (passionate wisdom) and "the belly for dividing of the stomach" (digesting food and absorbing through intestines). Next, he identifies the health-promoting calamus root which today is known to help heal a wide spectrum of ailments with the brain, nervous system, fever, gut issues, kidneys, and the liver. He goes on to associate the liver with wrath, gallbladder with bitterness, spleen with laughter, kidneys (reins) with craftiness, and loins with power. (227. Sharma V, 2014)

From what we've learned, it's easy to recognize the first two connections, where anger affecting the liver can eventually lead to gallbladder issues if not resolved and bitterness is allowed to take root. Conversely, a healthy liver may imply the ability to utilize explosive anger as a healthy survival mechanism which is then able to dissipate quickly without long-term effects.

The spleen's association with laughter is akin to "singing" in TCM, but not obvious from a current medical perspective. However, symptoms of spleen disfunction have been linked to depression and fatigue, so by inference, a healthy spleen may be able to impart a feeling of wellness including the ability to laugh more often in the absence of these symptoms. (228. Ma L, 2022)

At first glance, the craftiness of kidneys may appear to deviate from TCM's connection with fear, but looking from a different angle, craftiness could illustrate the ability to avoid risky situations by being discerning and shrewd; so, in that sense, it could be seen as an assertive approach to fear, in effect, short-circuiting anger and grief in the emotional cycle process by evading threats.

Lastly, the loins or groin don't have any physical power in and of themselves, but as current knowledge demonstrates, testosterone produced in the gonads is arguably the most powerful hormone in men responsible for promoting muscle development, youthful vigor, strength, and long-term health.

Given all these perspectives, it's remarkable to see how the sons of Jacob, separated by vast distance and centuries from TCM, exhibit amazingly similar insight into the knowledge of the human body and mind created by God and given to mankind. Divine power and insight are attributed to the success and exploits of the patriarchs, but undoubtedly, a healthy lifestyle was a key factor as well.

The Patriarch's Lifestyle and Diet – As wealthy semi-nomads, the Patriarchs, beginning with Abraham and his large household prospered within their own diverse economy and unique subculture.

Held as a princely people by those around them as, they possessed gold, silver, camels, livestock, deluxe household items, and servants among other things. Their pursuits included everything from plying as merchants, traders, pastoralists, and farmers; to capitalizing on spoils of war when push came to shove. (229. Caesar, 2006)

The size, affluence, and regard for their household from those around them however, appears to have waned from the time of Abraham to Jacob's twelve sons. By the time they departed for Egypt, their family had grown to seventy, but their servants and influence had diminished. Suffice to say, aside from a few years of famine toward the end of their time in Canaan, they probably lacked nothing in regards to a well-balanced and nutritious diet.

Although meat was less available for common Israelites later in history, it's probable that the patriarchs ate their fair share of meat, especially given their wealth status including all their flocks and herds. In addition, they had access to wild game as cited in biblical texts, which later became scarce for Israelites, so organic grass/field-fed red meat most likely played a role in their legendary vitality.

Nonetheless, one of their two biggest staples is evidenced by the copious amounts of dairy they likely consumed from all of the sheep, goats, cows, and camels they owned.

They ate dairy in various fermented forms such as sour milk, butter, and curds (cheese) as well as fresh milk. Like all dairy consuming cultures, this surely contributed to their strength and stature as they grew into adults. To a lesser extent, fish and wild poultry like quail, partridge, and other birds were part of the diet as well.
(230. YAHWEH's Sword, n.d.)

Aside from Zebulon's action of supplying fish during a famine, there are accounts of a dried fish trade occurring in the Levant during that time also. Fishing probably occurred primarily in the Sea of Galilee, but also to a smaller degree in the Mediterranean Sea since reliable ship technology had not yet been adopted.

Grains like barley and wheat for making bread were one of, if not *the* primary staples, along with beans, lentils, and chick-peas used in stews and other dishes. Vegetables likely included cucumbers, leeks, garlic, onions, radishes, and "bitter herbs" like lettuce, chicory, cress, parsley, and dandelion. Fruits included olives, figs, pomegranates, grapes, dates, and melons.

Pistachios, almonds, and wild honey were also available (Gen. 43:11), adding branch-chain amino acids (BCAAs for muscles), important minerals, and more antioxidants to their diet. Adjacent to trade routes in the Levant, the patriarchs would've had no problem obtaining salt as well as dates, especially being in close proximity to the Dead Sea and desert.

The significance of certain staples is seen when "The Lord" and two angels appear as men to Abraham outside his home in Genesis 18:6-8. Realizing their authority, he quickly runs and orders servants to prepare a calf, bake bread, and gather curds to present as guest food to the three men in what could arguably be the world's first recorded cheeseburger!....just saying.

On another note, these staples along with all the mediterranean fruits, vegetable, herbs, nuts, occasional fish and poultry, honey, salt, and well water provided the patriarchs with all the protein, BCAAs, healthy fats, carbs, antioxidants, and electrolytes they could possibly need to fuel their semi-nomadic lifestyle and campaigns as virtual Conans of the Levant and foregoing ancestors to the nation of Israel.

In a sense, they ate the ultimate Mediterranean or paleo-terranean diet. Some foods like dairy and meat, were scaled back (except for priests and wealthy) for common people centuries later, but though we cannot be certain of proportions, it surely served them well.

Ancient Israel Diet Overview – Many centuries later after Moses and Joshua led the Israelites back to conquer the land, the cultural diet and certain foods would become codified into the very identity and spiritual practices of Israel. The Israelite diet was probably nearly the same as their forefathers except for changes in proportions where meat and dairy may have decreased, while foods like bread, olive oil, and wine appear more prominent in the texts.

Shown below are a couple food pyramids contrasting the likely food priorities between the Patriarchs and their Israelite descendants. While it may not be precise with certainty, this is a best estimation gleaned from historical inferences and cryptic clues.

Figure 6. The Patriarchs Diet and Ancient Israel Diet

Of course, this varied according to region, occupation, wealth status, and other factors, but these are generalities concerning the common people. Folks near Galilee were likely able to eat more fish.

Regarding cultural diet in general, there are many correlations and groupings pointing to the significance of certain foods and staples whether by nutrition or spiritual importance. Modern monikers such as the bible diet, 7 biblical foods, foods of the feasts, etc. are examples of attempts to explore meaning and nutritional wisdom in the bible, and for good reason.

As most people are aware of these days, the mediterranean diet has been lauded as one of the wisest and well-balanced diets for health and longevity, so it's interesting to see how certain foods bridge ancient wisdom with modern research as we've already discovered.

Biblical Dietary Wisdom – Beginning with Moses, dietary laws were established to discern "clean" animals from "unclean" animals and other stipulations. One major criteria was the forbiddance of eating animals that do not both chew the cud and have cloven hooves (Deut. 14:7-8). For example, pigs have cloven hooves, but don't chew the cud, while rabbits chew the cud, but don't have cloven feet. Other prohibitions included shellfish and animals or livestock killed by wild animals.

On the other side of the equation, certain foods and spices were highly esteemed and used in holy rituals by priests during sacred festivals and days throughout the year. While the patriarchs were not technically kosher, biblical texts indicate that they already had a sense for clean and unclean animals and the prohibition of eating blood as evidenced by their diet and Noahide laws (Gen. 9:3-4).

Except for certain foods which later became either more significant or less available than others, the relevance of certain staples in the Patriarch's diet as emphasized in the Bible like bread, dairy, and meat, are underscored here in light of current research and health controversies, which we will address shortly.

Before conquering the land with Joshua, God promised to bring the people into "A land flowing with milk and honey" (multiple verses). Milk has already been historically considered by many cultures to build a strong people by supporting functions like bones, muscles, and immunity and Israel was certainly one of them.

The Israelites would have realized the maximum benefits from raw, grass-fed, A2 milk and fermented products as opposed to the pasteurized, grain-fed, A1 protein milk many folks consume today with diminished benefits, lactose intolerance and inflammation.

Moreover, current research has identified a fatty acid contained in milk fat called C:15 or Fatty 15 which promotes great benefits for long term health, in addition to a molecule called Neu5Ac necessary for cellular membranes. C:15 has not only been shown to improve metabolic parameters like cardio, immune, and liver health, but activates the AMPK longevity pathway as well. C:15 has also been compared to metformin and rapamycin as a natural alternative.
(231. Venn-Watson S, 2023)

As nature's sweetener and a compliment to milk, raw honey has been eaten since ancient times, but also used for medicinal purposes. Like many of the whole foods God has provided such as grain, dairy, meat, salt, and others; honey has suffered misconceptions from some in the health establishment. Far from being just another form of sugar, nothing could be farther from the truth.

Raw honey is a nutritional powerhouse containing compounds such as antioxidants and polyphenols, enzymes, some amino acids, B vitamins, minerals, and bee propolis; a substance having anti-fungal and anti-bacterial properties, good for the immune system. Raw honey provides a host of other benefits like brain health, wound healing, improved digestion, and energy. (232. Chin, 2024)

As if ancient divine wisdom was not enough, current research links within the article reference below corroborate the synergistic appeal of milk and honey from the bible as complimenting and boosting each other's health benefits. Both milk and honey together have been shown to boost immunity and relieve digestive disorders, improve quality of sleep, and support bone health. (233. Singh, 2024)

Moving on to another important food staple, Red meat, to varying degrees, continued to be an important diet component, but with a growing population, dwindling supply of wild game, and socio-economic factors, became more reserved for special events.

While over time meat consumption became more of a special occasion for the common folks during religious feasts, weddings, etc., the priests appear to have maintained a more consistent inclusion of meat in their diet because of ongoing sacrifices brought by people individually throughout the year for various offerings.

Contrary to modern perceptions of pious, reflective, and sedentary lifestyles for priests, the laborious duties and physical loads required of Israelite priests offering sacrifices was very demanding, so a higher meat diet would have supported their strength and vitality.

An interesting command from God in Leviticus reveals some intriguing insight that modern society would do well to take note of regarding the controversy over whether any animal fat is healthy to eat. The question is answered by not if, but what type is acceptable.

In Leviticus 7:23 God states "Give the following instructions to the people of Israel. You must never eat fat (or blood), whether from cattle, sheep, or goats", referring to bodily fat in other verses as well; but he never commands them to separate out dairy fat. While some unseen fat within the meat is normal, it's clear that they were to trim away all visible fat. (234. Nelte, 2016)

The main reason much of the red meat today is not heart healthy likely stems from its grain-fed high omega-6 content in addition to the excess saturated fat, marbling, and impurities it contains, as opposed to healthier grass-fed and fat-trimmed leaner meat with higher omega-3s and less impurities.

Nonetheless, legitimate concerns have been raised over red meat in general, regarding a sialic acid or acid sugar molecule found within red meat called Neu5Gc which is proinflammatory and may have implications for CVD and cancer. But like many claims today, it has more to do with how the animals are raised, their diet, the fat content, and balance of nutrition a person eats. (235. Patrick, 2025)

Case in point — a molecule related to Neu5Gc called Neu5Ac found in other foods acts as a competitive replacement for Neu5Gc and tends to regulate rather than aggravate the cells in blood vessels like Neu5Gc does. Research also shows that while meats like beef, lamb, and pork do contain Neu5Gc, beef fat for instance, contains three times the amount than the meat itself. (236. Failla S, 2023)

Furthermore, foods like dairy, eggs, poultry, and fish contain high amounts of Neu5Ac with little to no Neu5Gc at all. Inferring from all the research, this means that moderate amounts of grass-fed lean meat or even higher amounts along with more sizeable amounts of pasture-raised dairy, eggs, poultry and wild-caught fish is not only nutrient-dense and healthy, but non-inflammatory as well. This squares perfectly with what we know about the mediterranean diet and God's protective wisdom laid out in the bible. (237., 2014)

Research and common sense would therefore dictate that it is better to obtain necessary saturated fat from whole-fat dairy, used to nourish infant animals and humans at the beginning of life, rather than fat from meat which can over time, lead to poor outcomes.

Emerging as a dietary staple and culminating with the messianic decree of "The bread of life", bread played a central role in Israel. For millennia, people the world over have eaten cereal grains in the form of bread or similar foods and ancient Israel is no exception.

In fact, bread was not only a staple, but referenced many times to daily sustenance, in prayers, temple offerings, and even eternal life in the bible (2 Samuel 16:1, Matt. 6:11, Lev. 24:6, John 6:35). Moreover, unleavened bread was and still is central to religious feasts (Ex. 12:15, Luke 22:19) and bread is the most commonly mentioned food in the bible.

This begs the question: If bread is so detrimental to human health as many claim, why has it been a dietary staple for thousands of years and even self-referenced to Jesus himself as "the bread of life"?

Would God lead his people into a land abundant with milk, honey, red meat, and bread if these foods were unhealthy? The answer of course is no and has more to do with how these foods are raised, harvested, and prepared, not the food itself.

For example, glyphosate is routinely used on wheat crops, and while it may or may not be the primary driver of inflammation, it does appear to be a contributing factor. Celiac issues aside, the main problem going back 100 years is the separation out of the bran and germ from the endosperm leaving white processed flour which lacks nutrients, is more inflammatory, and has a higher glycemic index.

Furthermore, the bread of ancient Israel was nothing like "bread" found on American shelves today which is squishy, preservative-laden, and devoid of original nutrients. Bread in Israel was made from stone-ground wheat and barley mixed with water, salt, and wild yeast. Hence, it was fermented and nutritious, enhancing the gut biome rather than degrading it. (238. Barnett JA, 2020)

Of all the staples, Olive oil may have been the most revered staple in ancient Israel. Not only was it used in baking and cooking, but also in cosmetics, ointments, for lamps in the temple, anointing priests and rulers, and even currency. The spiritual significance of olive oil also underscores its health benefits. (239. Oestermann, 2021)

The nutritional properties of olive oil have been studied extensively and unlike other foods, olive oil is universally accepted as being very healthy and non-inflammatory. The many beneficial nutrient categories include fats, vitamins, antioxidants, and other compounds. At around plus or minus 75% of its total fat, the monounsaturated fat content of olive oil is one its greatest qualities.

These mono-fats along with vitamins E and K, potent polyphenols, and small amounts of phospholipids support olive oil as a nutritional and anti-inflammatory powerhouse improving heart and vascular health, cholesterol levels, blood pressure, dementia, and reportedly, even cancer. As far as Israel and many mediterranean cultures are concerned, olive oil is the elixir of life, but another celebrated liquid in Israel holds significance as well. (240. Fernando, 2023)

<u>Wine</u> in ancient Israel not only holds special meaning in terms of scripture and prophesy: i.e., "new wine", but has always been a commonly consumed beverage and to some extent, even considered a staple. While the bible warns many times against drunkenness, the overwhelming consensus is that moderate consumption of wine around meals and feasts was not only acceptable, but celebrated.

Even so, drinking more than a glass of wine a day still delivers a good amount of alcohol, so how did the Israelites appear to consume it so frequently without experiencing negative effects? Well, like many topics we've learned about, lost knowledge about ancient basic methods may hold the key.

According to research and ancient sources, the wine they drank had a much lower alcohol content. Actually, it wasn't the wine itself that was less potent, but the final drink. Historical sources explain that 11-12% wine was mixed with three parts of water before being served, reducing it down to about 3% alcohol. (241. Quarles, 2021)

This allowed for higher consumption during special events, while at the same time, promoting hydration and balancing the pros and cons of drinking undiluted wine. Other near-east cultures like the Greeks actually considered drinking straight wine barbaric, so it's interesting how modern society lost sight of this basic knowledge.

Wine contains many antioxidants and polyphenols, the most well-known of which is resveratrol. Thus far, the research here has shown that resveratrol is great for heart, kidney, lung, and brain health which basically covers most of the metabolically downstream organs, but it doesn't stop there.

As per research, resveratrol also promotes apoptosis, mitigates liver disease and fibrosis, improves insulin sensitivity, fat metabolism for weight loss, and longevity by activating sirtuins and the AMPK pathway. Red wine also contains quercetin which provides added antioxidant and anti-inflammatory properties. (242. Meng X, 2020)

In closing, the knowledge and insight of the Ancients is impressive. From ancient Chinese intuitions to divine biblical wisdom, we can see how in many ways, modern society is just beginning to catch up with and unlock the secrets of health from traditional wisdom.

Part 5 – Integrative Concepts and Health Modes

In this book so far, we've covered a lot of ground and information pertaining to a wide spectrum on health, vitality, and ancient wisdom supported by science and research. From snippets of my personal journey and practical tips to more in-depth science and ancient holistic insights, this information is hopefully inspiring, but I know is probably also a lot to digest and retain.

In Part 5, we'll circle back to summarize and synergize some of this knowledge as well as continue to introduce a few more concepts and health modes along with more graphics and diagrams to complete a fuller picture of integrative health. In this part of the book, there will be fewer cited references overall and more summarizing research on the research, so without further ado, let's jump in.

Summarizing Intro

When I experienced my health crisis three plus years ago, I knew something was very wrong — something had changed. It was more than a specific ailment, but rather a metabolic shift or maybe more accurately, a silent culmination that had reached a tipping point.

I'll share my vibrant recovery in more detail toward the end, but in a nutshell, I am doing much, much better now and in some ways, feel like I've turned back the clock quite a few years. As a result, I have more youthful energy, a sharper mind, and a new outlook on life, but progress didn't happen overnight.

In retrospect, I thank God my metabolic state wasn't beyond recovery, but you wouldn't have known it based on my symptoms. Everyone's experience is different, but I feel fortunate in that like an enemy attack on a fortress, my front line guard (gut & liver) recognized and took the brunt of the attack, however painful. The enemy even broke through to the middle guard (pancreas) for a time before being repelled by a drastic change in tactics, i.e., strategic diet and nutrients along with targeted acupuncture and Chinese herbs.

Sadly, many others suffer a sneak infiltration attack in an attempt to destroy the rear guard within the fortress as illustrated by years of asymptomatic and undetected gut and liver issues that can bypass the metabolic alarms and potentially cascade into more serious diseases like CVD, diabetes, KD, cancer, brain, and nervous disorders. That's not to discount other causes or genetics for these diseases, but just to say that lifestyle plays a huge factor in all of this.

Unfortunately, the processed food and pharma-influenced medical industries aren't doing us any favors. The medical establishment, via triage and modern science is good at putting Humpty Dumpty back together again, but not at preventing him from falling off the wall in the first place, so we need to be proactive with our health.

As we've seen from complex interactions between all the organs and biological processes, the human body is a brilliantly engineered control system design with countless feedback mechanisms necessary for promoting homeostasis or balance within the body.

Accordingly, the body is designed to thrive from all the natural food sources, traditional methods of preparation, and concentrated nutrients that modern tech can offer to help combat the environmental onslaught we face every day. Reducing stress, exercise, and taking advantage of various alternative treatments are also a great compliment for promoting a healthy lifestyle as well.

So with that, let's segue into the various sections that will help to inform targeted approaches, connect dots with graphical concepts, and link it all together. This first section is a bit chunkier, like in Part 2, but far too important to gloss over with just a basic review.

Metabolic Syndrome and Select Nutrient Flow Charts

Metabolic syndrome is one of the most chronic and pressing health issues of our day, so a special focus on metabolic processes relating to cholesterol, insulin resistance, CVD, and brain health are in order here — along with factors like LGCI (chronic inflammation) and autoimmune disorders which, according to research, can trigger these issues along with cancer, kidney disease, and others as detailed earlier in Part 2. (243. Grivennikov SI, 2010) (244. Christovich A, 2022)

LGCI is largely triggered by autoimmunity, which in turn is mainly caused from intestinal issues like impaired gut biome and leaky gut, which in turn again, can be altered via poor diet and unhealthy bile from the liver that can damage the tight junctions of gut lining by allowing harmful bacteria to proliferate over good bacteria.
(245. Guzior DV, 2021) (246. Dmytriv TR, 2024)

Back in Part 3, a couple of complex, but foundational flow chart diagrams were illustrated, detailing major biosynthesis pathways for amino acids, their metabolites, some of the major hormones, and energy production in cells. The idea was to give a comprehensive overview of how all the nutrients contribute to healthy metabolism.

On the following pages, we'll investigate a couple of slightly less complicated diagrams along with a few more simplified flow charts. The upper diagram of the first figure illustrates how healthy cholesterol and bile along with "browning" of fat and the roles of taurine, glycine, choline, and other crucial nutrients play in converging together to produce metabolic health.

Cholesterol Metabolism – The upper diagram in Figure 7, details how cholesterol management is governed by healthy bile acids. Recall in Part 2 how the FXR receptor in liver cells regulate bile acids, but here we see how FXR is also regulated by bile acids via probiotics and taurine in a feedback loop. (247. Wolska A, 2018)

FXR along with thyroid hormone T3 triggers the APOC2 gene, which in turn along with omega-3s and BAT advances healthy triglyceride metabolism, promoting large buoyant LDL (lbLDL) over atherosclerotic small dense LDL (sdLDL). (248. Froyen, 2021)

WAT and BAT are two forms of body fat that refer to white adipose tissue and brown adipose tissue, respectively, and play a significant role in metabolic health. Also note the foundational roles methionine and SAM play below that, which will be explained later.

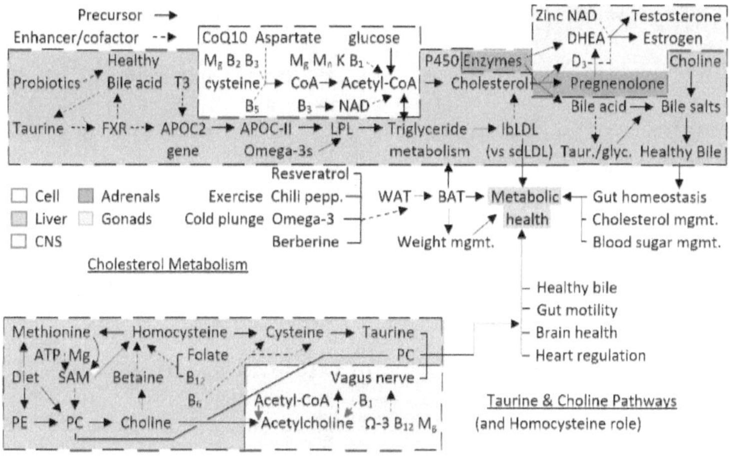

Figure 7. Pathways for Good Cholesterol, Bile, & Healthy Outcomes

White adipose tissue or WAT, which includes subcutaneous and unhealthy visceral fat, is transformed to the healthier brown adipose tissue or BAT through the various mechanisms listed above like exercise, cold plunging, resveratrol, chili peppers, omega-3s, and berberine to name a few. Through a process called thermogenesis, BAT is very efficient at burning or metabolizing triglycerides as opposed to WAT which stores them, leading to better health and weight management. (249. Hughes, 2022)

The upper diagram also reveals how the main pathways to acetyl-CoA from glucose, triglycerides, and amino acids along with micronutrients promote healthy cholesterol metabolism for bile acid and hormone production; however, excess cholesterol not excreted or used can accrue in the liver and also blood vessel walls via sdLDL.

Following acetyl-CoA, P450 enzymes merge with cholesterol to produce functional hormones and bile acids. Taurine and/or glycine conjugates with bile acids to produce bile salts which then incorporate choline at the end of the process for healthy bile, excess fat removal, and proper detoxification capacity in the liver.

Excess cholesterol coupled with sdLDL, otherwise known as small dense LDL cholesterol, can penetrate and lodge into blood vessel walls, but permeability is much greater with damaged vessels than with healthy vessels. Special cells within the endothelial blood vessel walls called macrophages encapsulate this cholesterol when triggered by inflammatory cytokines to form what are called foam cells, which if left to grow and accumulate, can lead to plaque lesions and atherosclerosis or hardening of the arteries, or worse, a rupture.

The next step to help mitigate this disease, besides adequate T3, omega-3, taurine, glycine, choline, and BAT for healthy cholesterol and bile formation, is to incorporate certain other nutrients targeted by research to improve atherosclerosis. While nutrients like allicin (from garlic), curcumin, CoQ10, and vitamin C help mitigate CVD by improving inflammation and endothelial function, others like resveratrol and berberine may actually help to prevent or even reverse atherosclerosis according to research. (250. Moss JW, 2016)

It's interesting to note how some of these nutrients overlap with those used to convert WAT to BAT, so with that, let's take a look at some simpler diagrams to summarize what we've covered so far.

The first flow chart below summarizes the acetyl-CoA-cholesterol pathway and the nutrients, including P450 enzyme nutrients, that assist with promoting robust hormones and functional bile, along with healthier gut, cholesterol management, and insulin sensitivity.

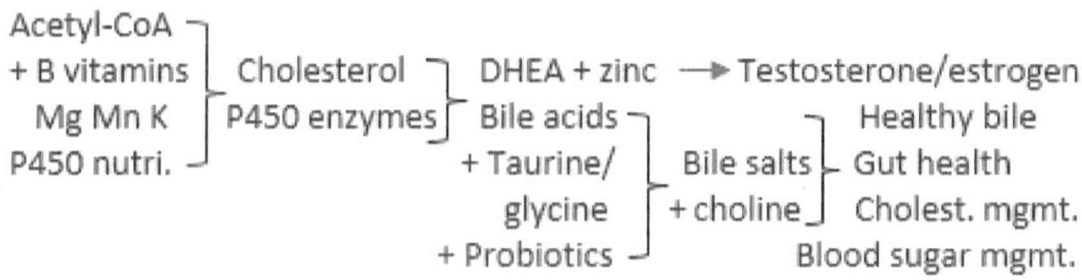

P450 Enzyme Nutrients

iron
B-vitamin complex
cruciferous flavonoids
resveratrol
zinc
magnesium
vitamin C

Figure 8. Nutrient Pathways for Healthy Hormone & Bile Metabolism

The second flow chart distills the Taurine-APOC2-triglyceride pathway and nutrient cofactors that promote lbLDL cholesterol and a lower risk of atherosclerosis.

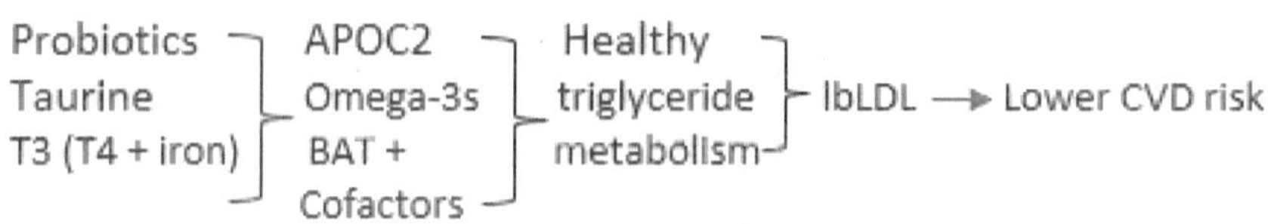

Figure 9. Nutrient Pathway for Healthy lbLDL Cholesterol Management

Between both charts, we can see how crucial factors like probiotics, taurine, omega-3s, glycine, T3, choline, B vitamins, minerals, BAT P450 nutrients, and exercise are in promoting healthy cholesterol, hormones, and bile leading to a whole range of health outcomes.

Taurine and Choline Pathways – Moving on to the lower diagram in Figure 7, the chart illustrates different, but related pathways also leading to better health. Whereas taurine and acetyl-CoA initiate the processes just discussed, methionine and PE or PC initiate the ones in the lower diagram via diet. Homocysteine, which is a direct metabolite of S-Adenosyl methionine or SAM, has garnered much attention in recent years due to the prevalence of MTHFR gene defects which impair methylation in up to 40% of the population.

Methylation is the process by which a methyl group containing three hydrogen and one carbon atoms attach to and regulate various molecules such as DNA, RNA, proteins, phospholipids, and neurotransmitters via gene expression, cellular membrane integrity, neurohormone synthesis (serotonin, dopamine, epinephrine), red blood cells, creatine, and amino acids. (251. McCaddon A, 2023)

In the context of Figure 7, the essential AA methionine requires ATP and magnesium to biosynthesize SAM, which plays a central role in this chart. SAM then donates a methyl group for the various processes listed above while simultaneously being catabolized to form homocysteine, which can then be biosynthesized into cysteine with the aid of vitamin B6, or converted directly back to methionine using either the betaine or folate-B12 pathways.

SAM acts to regulate homocysteine by enhancing either the production of cysteine when methionine levels are high or by converting homocysteine back to methionine when levels are low.

It's interesting to note that PE and SAM team up to biosynthesize PC, which in turn biosynthesizes betaine via choline if dietary PC is not available, thus forming a feedback loop for remethylation. So, what does all this mean for those with the MTHFR gene defect, which causes elevated homocysteine levels from the inability to remethylate properly via the standard folate-B12 pathway?

Well, despite decades of research, there isn't widespread consensus on clear-cut outcomes for elevated homocysteine levels in academia. However, the two most plausible conditions linked with MTHFR in research appear to be limited detoxification capacity in the liver via reduced cysteine, choline, and glutathione – a metabolite of cysteine – and a lack of SAM-inducing anti-inflammation in blood vessel macrophages which may lead to dangerous plaque legions and higher risk of CVD. (252. Pfalzer AC, 2014)

Moreover, evidence shows that a metabolite of choline and carnitine from bad bacteria called TMAO can contribute to atherosclerosis, but this can likely be mediated by simply eating fiber along with fermented foods or probiotic supplements. (253. Liu Y, 2020)

Summarizing this section, consuming dietary PE, PC/choline, and probiotics to support the betaine pathway along with plenty of B6 to support the cysteine pathway leading to taurine and glutathione synthesis can help compensate for impaired MTHFR methylation, aside from taking methyl-folate and methyl-B12 supplements. The simplified flow chart next summarizes this portion of the diagram.

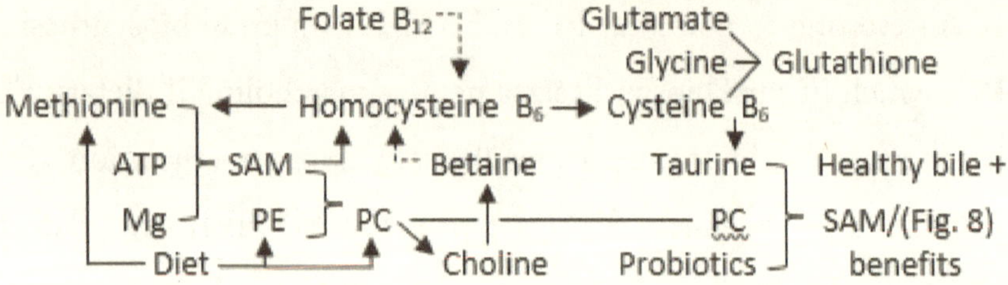

Figure 10. Nutrient Pathways for Methylation, Cysteine, & Choline Pathways

As one can see, SAM is critical in regulating homocysteine in both directions, helping to biosynthesize non-dietary choline and PC via PE, supporting cysteine and taurine synthesis, and in the process donating methyl groups for a whole slew of vital functions including DNA, neurotransmitters, phospholipids, red blood cells, proteins, and creatine; with methionine and magnesium as the foundation.

The final breakout chart for this section features the lower portion of the lower diagram in Figure 7 which illustrates the common sources of PC and choline, along with other nutrients such as B_1, omega-3s, B12, and magnesium; leading to healthy vagus nerve function. The vagus nerve is a crucial communication pathway for the CNS between the brain, gut, and other organs influencing functions like gut motility, cognitive function, heart regulation, mood, and inflammation among others.

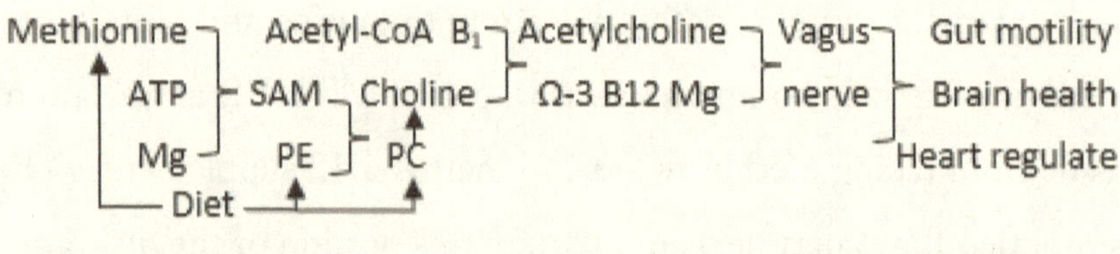

Figure 11. Nutrient Pathway for Choline and Vagus Nerve Regulation

Though somewhat focused, the nutrient pathways illustrated in Figure 7 and breakout charts 8-11 encompass an array of processes central to key metabolic health issues people experience today, with the exception of some of the endocrine and kidney hormone paths included in the more comprehensive charts back in Part 3. Even so, as a partial endocrine organ, the pancreas also benefits from better insulin regulation via healthy bile which controls blood sugar.

As we've seen many times now, liver and gut dysfunctions, and by extension, central nervous system impairment can damage other organs downstream. Thus, healthy gut biota along with functional amino acid, enzyme, cholesterol, bile, and hormone biosynthesis is absolutely critical for metabolic health with the liver playing an extremely central and dominant role in all of this.

Nutrients like probiotics, T3, taurine, glycine, omega-3s, PE, PC/choline, methionine, cysteine, B vitamins, magnesium, manganese, zinc, potassium, P450 nutrients, and resveratrol, are among the most crucial cofactors for sustaining metabolic health in this context, but other factors like exercise, reducing stress, alternative treatments, and obtaining proper sleep are vital as well.

Finally, bi-directional cross-talk between the brain and gut via vagus and splanchnic nerves serve not only as vital feedback signals for metabolism, but can also trigger symptoms as early warning signs that something is off with the liver-gut-brain axis and that we should pay more attention to our health amidst life's distractions.

The Great Cheeseburger Conundrum

Everyone enjoys the taste of a great cheeseburger whether they can dietarily tolerate one or not, but unfortunately the way processed food is today, not everyone can or wants to eat them. Underneath the condiments and sparse veggies, a cheeseburger basically consists of bread, meat, and cheese, if it can even pass for that.

The ingredients from three of these basic food groups has been the topic of great debate within the health and medical communities for many years now; some claims valid, some misunderstood, and some just wrong. In the Patriarch section of Part 4, I joked that Abraham presented to The Lord and two angels the world's first cheeseburger because the meal consisted of bread, curds (cheese), and meat.

The fact that Abraham presented this meal in haste and reverence underscores not only his great respect for God, but also that these foods are not inherently bad, but actually very healthy when raised, fermented, and prepared the right way as part of a healthy diet. The case for healthy dairy and meat was largely made in Part 4, and to some extent for bread, but a little more needs to be said on bread.

Bread – Amidst the noise and confusion, there aren't very many nutritional experts on bread, but Sue Becker from breadbeckers.com is truly one of them. With a degree in food science and decades of experience and teaching, she cuts through the hype to address hot topics like phytic acid, sprouting grains, natural leavens, and the amazing health benefits fresh-milled whole grain bread has to offer.

In her articles, she points to inspired testimonials from others of how switching to fresh-milled grain products has alleviated all sorts of ailments related to gut issues, cholesterol, blood sugar, and overall detox, in effect, eliminating conditions like skin disorders, chronic headaches, allergies, and infections. (254. Beckers, Bread, 2022)

Moreover, a gut enzyme called phytase which breaks down phytic acid is shown to bind with radioactive and toxic heavy metals in order to remove them from the body. Cholesterol and blood sugar levels are likely mediated by PC and insulin-regulating inositol from eating whole grains, but other factors like fiber contribute also.

Sue also points to research showing how, although studies conclude that phytic acid reduces nutrient absorption in a certain context, the comprehensive picture actually reveals how the gut can adapt to produce more phytase via phytic acid signaling for better absorption of minerals over time, once whole grains are introduced. Maintaining a healthy and diverse diet can also negate concerns over losing some of the minerals to phytic acid. (255. Becker, 2017)

Aside from natural acclamation in the gut, sprouting grains and fermenting dough via wild sourdough yeast can also reduce gluten and phytic acid for those who are more sensitive or intolerant to grains, while at the same time, upregulating certain nutrients. However, this may have the unintended result of lesser benefits for cholesterol and blood sugar control, as well as carcinogenic protection in the colon, so a balanced approach may be the way.

While all wheat has certain amounts of macro and micro-nutrients, Kamut and other ancient grains have higher amounts. One source lists Kamut as containing 10 grams of protein and 8 grams of fiber per cup as well as 28% DV of zinc, 20% each of magnesium and phosphorus, 13% of B1, 33% of niacin, 77% of manganese, 40% of copper, and nearly 100% of selenium. Not bad for unfortified bread! This gives "bread of life" a whole new meaning. (256. Frey, 2021)

Dairy – You can't have a great cheeseburger without cheese, so the next ingredient up for review is dairy. In Part 4, the nutritional case was made about how organic, grass-fed, A2 dairy, especially in its fermented form, is not only very nutritional, but anti-inflammatory as well. In addition to C:15 fatty acid for metabolic health and Neu5Ac for crowding out meat-derived Neu5Gc in cellular membranes, dairy also contains protein, calcium, potassium, B_{12}, and higher omega-3 content in grass-fed dairy. (257. Devje, 2022)

Regarding fermented dairy, research continues to confirm its superb health properties. These include intact probiotics along with postbiotic metabolites like SCFAs and antimicrobial peptides which help to enhance the mucosal barrier and tight junctions of the gut lining in response to conditions like leaky gut, immune issues, and malformed bile acids. (258. Gao Y, 2025)

This helps to improve metabolic parameters like LDL cholesterol, blood pressure, obesity, diabetes, and cardiovascular health. Even so, it's also helpful to view benefits through the lens of culture.

As mentioned in Part 4, nearly all longevity hotspots like the blue zones and Caucasus, except for Okinawans who eat fermented vegetables, consume fermented dairy. The healthy long lives of these people bear testament to and confirms the scientific validity of how fermented foods, and in this case fermented dairy, alongside a balanced whole-food diet, help to increase longevity by promoting healthy gut biomes which affects everything else downstream in a positive way. (259. Mohol P, 2025)

Meat – No matter how good the quality of bread and cheese, you can't have a cheeseburger without meat. Again, the case for lean grass-fed red meat alongside a healthy balanced diet has already been made; including higher omega-3s, a better protein profile, reduced Neu5Gc via trimmed fat and competition from Neu5Ac in other foods, and decreased TMAO production via fermented and fiber-rich foods. However, there is one more key factor to consider.

One final aspect that affects the beneficial outcome of consuming meat has to do with how it is cooked. Advanced glycation end products or AGEs are harmful by-products formed between sugars and proteins or fats that exist in foods to varying degrees.

A rating system has been established to gage the amount of AGEs formed from a particular method of cooking and/or combining foods, that are especially high in meats. The outcomes for oxidative stress and inflammation caused by AGEs include diabetes, kidney disease, cardiovascular disease and cancer. (260. Uribarri J, 2010)

AGE content is directly related to the amount of heat and moisture used in cooking. For instance, frying, grilling, and roasting, produce much higher AGE content than boiling, poaching, and stewing. Another method that allows you to enjoy the best of both worlds is to marinate meat in lemon juice or vinegar before roasting or grilling to lower AGEs, although by not as much as boiling or stewing does.

An interesting account in 1 Samuel 2:16 describes how rogue priests and their assistants would threaten patrons bringing sacrificed meat to be boiled in a pot if they didn't hand it over to be roasted instead. 'The man offering the sacrifice might reply, "Take as much as you want, but the fat must be burned first." Then the servant would demand, "No, give it to me now, or I'll take it by force."'

At first glance you might be thinking 'Hey, can you blame a guy for wanting BBQ?' Yet, this action was in direct violation of God's command for priests to boil the fat off first. Not only do we see the discarding of fat as ritual obedience, but we also see God's wisdom with the principle of AGEs in protecting the health of his priests. So we see that the healthiest form of meat is grass-fed, fat-trimmed, boiled/stewed meat alongside a balanced whole food fiber-rich diet.

In summary, the cheeseburger analogy in its healthy form appears to be a perfectly fine combination of foods most people can enjoy. Of course, this in no way discounts the incredible benefits fruits, nuts, vegetables, legumes, fish, and poultry, have on our health, but this just adds a more balanced perspective to the overall equation.

Holistic Engineering and the Metabolic Organ Pyramid

Progressing through this study, we see the ever increasing evidence of an amazing and brilliant systems engineering design of the human body unfolding, as shown by the innumerable biochemical and bioelectrical processes imbedded withing the various organs, redundant backup systems and complex feedback cycles that all communicate with each other either directly or indirectly.

As such, all the organs are linked together metabolically with an apparent order of downstream effects. Part 2, lists all the organs from the bottom up, starting with those most vital for immediate survival and/or which tend to have the most severe consequences in later stages of life. Here, we'll review organs from the top down.

Chicken-or-the-Egg (Liver or Gut) – At the heart of this holistic system lies the gut-liver-brain axis with the gut and liver as the nutritional and hormonal drivers of this trio, but also gatekeepers of toxic relief. In response, the brain via NS, reciprocates by sending regulating feedback to these organs, but it can also directly initiate gut disturbance via the limbic system from chronic stress and worry.

To recap one of the main themes in the Introduction, pathogenic conditions start with the gut, but impaired symptoms that trigger the metabolic cascade likely begin with the liver because of its vast regulatory impact on the entire body. Some experts may disagree given the iterative feedback between both organs, but to get a better grasp, let's take a closer look at this chicken-or-the-egg enigma.

In Part 2, the G.I. Tract and Liver sections described how ideally, primary bile acids — which are deconjugated or separated in the gut by good bacteria — recycle back to the liver for reconjugation with primarily taurine, but also glycine, before picking up choline and returning to the gallbladder to be secreted into the duodenum.

Apart from direct gut disruption from stress via chronically elevated cortisol and adrenaline, the liver appears able to buffer the effects of bad gut microbes as long as taurine-conjugated and choline-infused bile can support the good bacteria's dominance and ability to deconjugate primary acids into secondary acids which help destroy bad microbes and protect the integrity of the gut lining.

However, once the balance shifts over time, the primary bile acids' ability to activate FXR receptors to upregulate taurine is diminished as bad gut microbes suppress their ability to deconjugate into the secondary bile acids which normally control them. (261. Ni Y, 2022)

This leads to a dominance of harmful gut bacteria which disrupts the gut lining and dysregulates bile synthesis, harming both the gut and liver in a vicious feedback loop. Moreover, TMAO metabolized from excess choline and carnitine by this harmful bacteria increases the risk of cardiovascular disease. (262. Anand S, 2022)

The resulting disfunction leaves the liver more exposed to reduced detoxification capacity, fatty liver and other liver ailments, but also excess bile buildup as a compensation factor for the inability to deconjugate primary bile acids in the gut.

This can lead to bile sludge and/or gallstones causing further disruption to the gut with issues ranging from asymptomatic leaky gut in the early stages and throughout, to disorders like SIBO, IBS, and even bile reflux in the form of gastritis and GERD. Moreover, congested bile acids can leak into circulation and accumulate in the hypothalamus, disrupting the blood-brain-barrier and HPA axis.

If left unchecked, these conditions may lead to further downstream issues or metabolic diseases in other organs such as the adrenals, thyroid, pancreas, kidneys, heart, lungs, and brain, along with other bodily tissues. So, in contemplating the origin that sets this process into motion, which is it, the liver or gut? I guess the answer is yes, but in my opinion, it's really the liver because it helps regulate the gut until bad microbes or high cortisol impair its ability to do so.

The Metabolic Organ Pyramid – The gut-liver-brain (HPA) axis forms the basis from which all lifestyle-induced pathologies begin before progressing to increasingly more severe metabolic organ consequences down the line. Communication between these organs is bi-directional via biochemical and vagus nerve signaling.

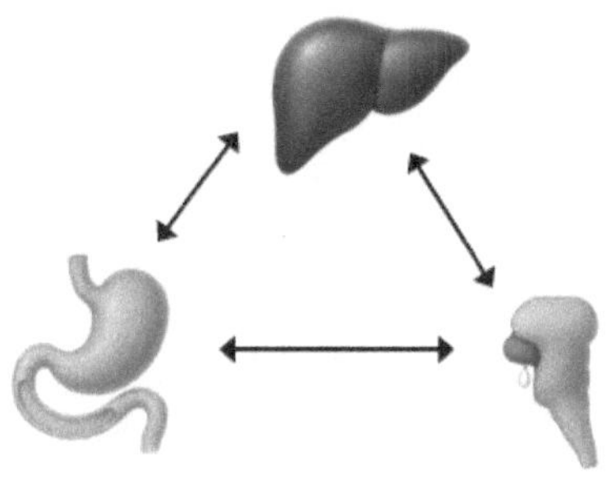

Figure 12. Gut-Liver-Brain Axis-Triangle

Once the gut-liver-brain axis or triangle is damaged, the domino effect of metabolic disease to other organs is only a matter of time if not corrected, but many people are asymptomatic until serious diseases arise. Summarizing the research, here are a few examples of how this metabolic course can cascade from one organ to another:

- The liver, via poor diet or trauma/stress-induced cortisol, leads to an impaired gut, BBB, and HPA axis; all of which in turn, harm the liver even more by reducing detoxification, bile/amino/hormone synthesis, and nutrient availability
- The liver, via altered gut biome, leaky gut, and HPA axis disfunction can impair the adrenal, thyroid, and gallbladder functions leading to further disruptions downstream.
- Consequently, the pancreas, kidneys, thymus, and spleen can be disrupted and/or diseased in the form of diabetes, enzyme deficiency, kidney disease, and immune issues
- Over time, the impairment of these organs along with the liver and gut, can lead to CVD, lung, bone, and brain disease

This slow motion metabolic cascade is best illustrated by the next chart in what I like to call the metabolic organ pyramid or MOP. For the purposes of simplicity, many of the feedback loops and nuances are omitted, but the pyramid basically reveals the metabolic downstream effects of organs on each other induced by lifestyle and aging, from top to bottom, but also in time from left to right. It may not be perfect, but I believe it to be fairly accurate given the research.

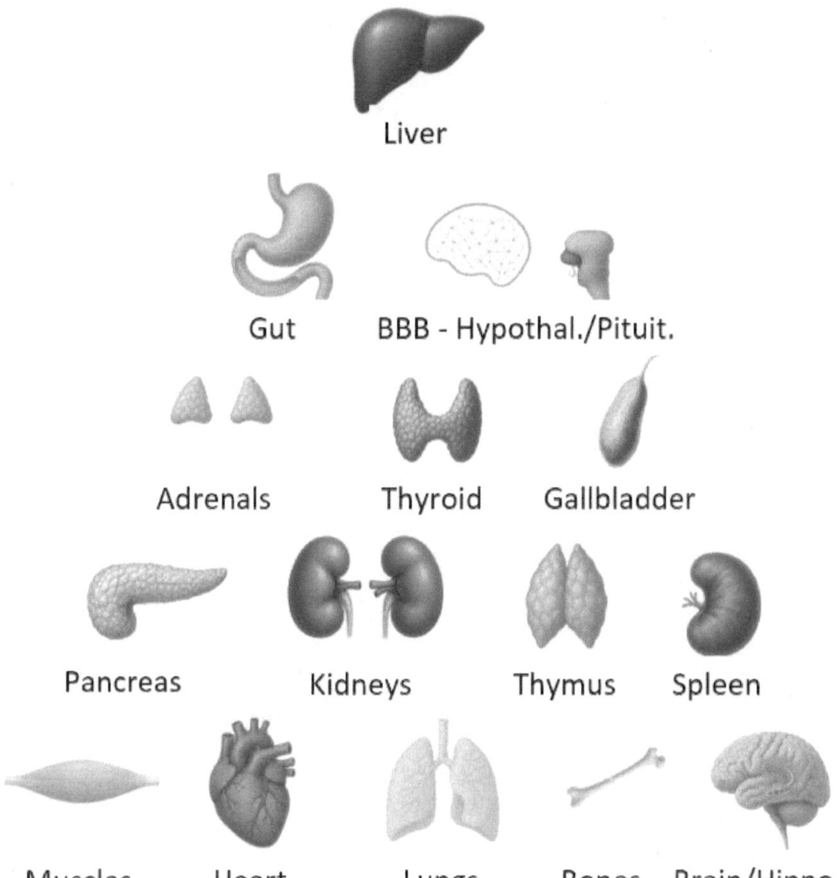

Figure 13. Metabolic Organ Pyramid (MOP)

Here, we see the liver atop the pyramid as king. Again, this order does not preclude genetic disorders which can predispose a more direct path between organs, or ultimate control because otherwise the brain would be on top; but rather, illustrates the general pathways to metabolic diseases over stages of life. The good news is that we have a roadmap to try and begin reversing disease over time.

To be clear, however, this does not replace the need for medical consultation when it arises, especially for more severe conditions, so whether by naturopathic, conventional, or functional medicine, professional help should always be sought out to make sure all the bases are covered, in addition to enacting healthier lifestyle choices.

Even so, on the flip side of this consequential MOP chart, we can begin targeting organs nutritionally based on where we're at metabolically; not only by addressing affected downstream organs, but also by starting at the top of the pyramid where it all begins.

Just as metabolic consequences flow down, so too can the benefits of preserving a healthy liver and gut, though it may take some time. In Part 6, we'll explore summary conclusions on organs, ailments, nutrients, and food, but for now, let's consider a few more holistic reveals and historical clues to reinforce the increasingly synergistic evidence linking traditional wisdom and current science.

More Traditional Wisdom – From ancient times till now, research continues to validate ancestral insights once thought to be mistaken. Part 4 and the sections just covered have bridged many of these links between ancient wisdom and modern science including TCM organ theory, alternative treatments, and biblical dietary wisdom.

In fact, we can see the TCM organ control cycle back in Figure 4 imbedded within the MOP chart going down the pyramid, not to mention all of the insights gleaned from the paleo-terranean diet consumed by the Patriarchs which have been confirmed by science.

Once more, the liver's status has always held a very prominent place in human history from ancient times right up until current and growing societal awareness. Intriguing perspectives esteeming the liver, as noted in historical Middle East records, detail ancient sentiments about this regarded organ. We already know the liver is negatively associated with anger, but much like how historically, the heart has been associated with passion and desire, the liver too also holds positive and grounding attributes in Eastern culture.

In Hebrew, the word for liver is derived from the three letter root K-B-D that also forms the words for important, honor, majesty, and glory, but also interior and soul. According to tradition in the Middle East, the liver conveys courage, perseverance, and desire, but is also used as a term for endearment as in "you are my liver", akin to "you are my sweetheart". Lamentations 2:11, also equates "liver poured out on the earth" with a state of spiritual or physical collapse in the absence of these attributes. (263. Reuben, 2024)

Whether by ancient adages or new novel discoveries, we can observe intriguing connections between East and West, but the correlations don't stop with organs. We can also see a relationship between ancient TCM terminology and current scientific jargon as it relates to metabolic and energetic pathways in the body.

Case in point is illustrated by similar inferences from both schools of thought regarding the interactions and effects between various foods, nutrients, and metabolic processes within the body.

For instance, in traditional Chinese medicine, TCM describes yang energy as warming, active, and energetic whereas yin energy is considered cooling, passive, and restorative. From a current research perspective, this sounds a lot like the contrast between not only the sympathetic and parasympathetic nervous systems, but also between mTOR and AMPK enzyme pathways in the body.

Recall that in Part 2, mTOR supports new cell and tissue growth along with mitochondrial energy via ATP, while AMPK promotes autophagy or cell repair, energy balance, and oxidative stress relief. In short, mTOR correlates to anabolic pathways and AMPK to the catabolic pathways. So here we can observe another link between East and West with yang correlating to anabolic processes and yin with the catabolic processes as shown in the graphic below.

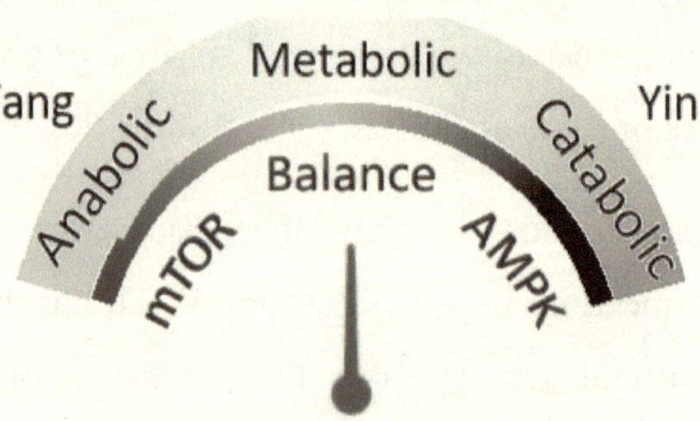

Figure 14. Metabolic Spectrum Guage

The body needs a balance of mTOR and AMPK, but the needle tends to skew left or right depending on life stage or health status. In the next section, we'll explore the wide ranging aspects and impacts of exercise, mTOR and AMPK on metabolic health.

Health Impacts of Exercise Via mTOR and AMPK Pathways

Taken as proxies for anabolic and catabolic pathways, mTOR and AMPK are regulated primarily by factors such as age, nutrition, exercise, and eating schedules, but are also affected by stress, sleep, and circadian rhythm. Here, we'll survey dynamics between exercise and metabolism, but specific routines will be left to fitness trainers.

Whereas, an mTOR dominant regime can reflect childhood growth and healthy adult athletic metabolism, it can also negatively reveal unhealthy tissue like senescent cells and tumors. Conversely, an AMPK dominant path can positively reflect enhanced metabolism, fat burning, cellular regeneration or autophagy, and longevity, but if too powerful, can inhibit muscle and stem cell repair and also suppress growth in children and adolescents.

Foods typically associated with the mTOR path include high protein foods like meat, eggs, dairy, fish, poultry, etc. and more simple carbs like starch vegetables, processed flour, and sugar. AMPK associated foods include polyphenol foods, fiber-rich vegetables, healthy plant fats, fermented foods, and spices. Healthy fats can support both pathways, but probably AMPK more so.

Exercise is also a potent activator between the two pathways. While there is some overlap, resistance exercise (RE) primarily stimulates mTOR while endurance exercise (EE) primarily activates AMPK. High intensity interval training or HIIT is a hybrid type of exercise that blends elements of both resistance and endurance training.

Younger people can generally thrive on all types of exercises, but so can older adults with some caveats. Whereas RE can be challenging for older adults without adequate protein and muscle recovery times, EE can likewise be too catabolic for seniors if their lean build and lifestyle already skew too heavily toward AMPK to begin with.

A research review shows that while older adults can blend RE and EE together, the timing between workouts is more critical to ensure that each exercise type doesn't cancel out the benefits of the other. Simply put, there should be ample time for muscle recovery after an RE workout before a more catabolic EE workout, but conversely, enough mitochondrial energy renewed after an EE grind to support muscle repair for an ensuing RE session. (264. Hajj-Boutros G, 2022)

HIIT training, which incorporates short bursts of intense exercise in intervals, is a balanced approach which according to research, can eliminate the timing issue and is a very effective activator of both mTOR and AMPK pathways for young and old alike. Exercise that would otherwise be very high rep, lightweight motion is accelerated into very fast thrusting movements which activates mTOR via light pseudo-resistance training while at the same time triggering AMPK and providing a cardio workout from the intense activity.

The benefits from these types of exercises depend on many variables such as age, mobility, and level of fitness, but while HIIT may appear to be the most ideal, the research shows it may not be the most optimal form of exercise in isolation depending on context.

HIIT alone can work for those of all ages with time constraints, those wishing to optimize cardiorespiratory fitness over muscle mass and strength, and older people without major mobility issues. While efficient and balanced in approach, HIIT may still fall short.

EE alone may be appropriate to all ages for weight management, cardiovascular health, and those recovering from injuries or surgery, but adding in context-dependent RE can enhance results. RE alone can be sufficient for younger adults in good metabolic health, but probably not for older adults because of the lack of cardioprotective health benefits from not including EE or HIIT on some level.

The research suggests that while younger adults can benefit from virtually any blend of exercise depending on goals and health, the optimal choice for older adults, subject to mobility is RE plus HIIT. This approach supports muscle mass and strength via RE and mildly with HIIT, while also promoting autophagy and antioxidant pathways via AMPK through the cardiorespiratory aspect of HIIT, without a substantial depletion of mitochondrial energy that results from EE. Even so, there is one more potent option to consider.

If HIIT is metabolically in between EE and RE, then circuit resistance exercise or CRE is between HIIT and RE. CRE, which involves fast-cycled, multi-targeted, moderate weight training as opposed to traditional RE focus on certain muscle groups at heavier weights, retains the benefits of muscle building while still providing some cardiovascular exercise, albeit at a lower level. (265. Hu C, 2024)

Still, yet another approach, especially for untrained heavier adults may be to build a cardiorespiratory foundation first by starting out with light to moderate EE, then shifting into vigorous EE (VEE) (metabolically between EE and HIIT), before transitioning into a more cycled exercise routine including HIIT and RE. Although strenuous, VEE and HIIT enhance maximum oxygen use (VO2 max), mitochondrial function, and lactate clearance, delaying fatigue during RE workouts and speeding recovery between sets.

Research also supports this concept in a daily form with the fine tuning of RE plus HIIT combo, but with slightly fewer sets within the same day and in reverse order. Within the context of a weekly routine, intraday HIIT followed by RE has been shown to improve cardio-metabolic parameters while enhancing muscular fitness and body composition faster than other workouts, but it can be intense, so sets should be tailored to the person. The graphic below illustrates how each type of exercise lies on the metabolic spectrum from anabolic to catabolic. (266. Nuñez TP, 2020) (267. Green, 2018)

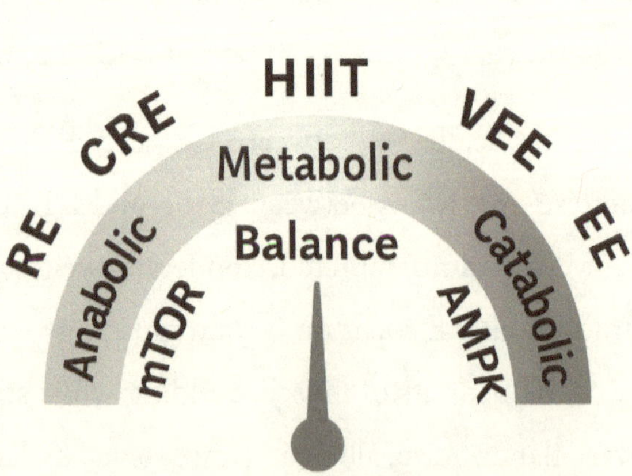

Figure 15. Exercise Spectrum Guage

The many variations of exercise in different combinations appear to be endless, but the facts are clear that exercise in its many forms are potent activators of both mTOR and AMPK depending on workout type. Proper nutrition is key to healthy stimulation of mTOR and AMPK, but exercise may be just as or even more important for activation, so proper guidance should be considered.

In addition to nutrition and optimal exercise, other factors such as stress, sleep, and alternative treatments, also play a crucial role in pursuing holistic health, so with that, let's segue to the next section.

Stress and Sleep

Aside from the two biggest factors of nutrition and exercise, comprehensive health depends on other crucial aspects like sleep and how we deal with stress. Quality sleep and a stable circadian rhythm help to buffer stress, both of which can be tuned by diet and daily activity. Apart from modern disruptions like chemical toxins, EMF, and off-rhythm blue light, health basically comes down to nutrition, exercise, sleep and ability to cope with emotional stress.

Stress - Everyone knows that stress can have a negative impact on health, but aside from general awareness of its effect on gut function and heart health, many sources can be fairly vague on the subject. Although acute stress can impact the vasodilation of blood vessels, most CVD aspects from stress appear to be indirectly driven by chronic long term issues upstream from the liver and gut or present in people already coping with CVD. (268. Anand S, 2022)

We've already seen, both in TCM and science, how worry and stress can trigger the gut directly via the nervous system, but exactly how does persistently high cortisol resulting from unresolved trauma, anger, and chronically elevated fight-or-flight modes downgrade the liver? Well, research affirms not only this pathway, but also the mechanisms in the liver by which it occurs. (269. Joung JY, 2019)

Whether by stress induced cortisol or altered gut biome via bad diet, a damaged gut barrier caused by either path allows a toxin called lipopolysaccharide (LPS) to enter the blood and disrupt cells in the liver called Kupffer cells which happen to make up about 15% of liver cells as opposed to hepatocytes which make up about 60%.

Kupffer cells are effectively the liver's recycle plant that, in addition to degrading cellular and microbial debris along with pathogens and toxins, also metabolize lipids, cholesterol, and red blood cells. When excess LPS (a shredded component of bacterial membranes) escapes into circulation through a leaky gut, it can damage Kupffer cells, triggering inflammation and fat accumulation in the liver.

Concurrently, stress reduces blood flow to the liver causing hypoxia which piles onto the damage LPS creates. Thus, chronic stressors like anger, trauma, and PTSD may actually be more damaging to the liver than diet in the shorter term via both of these pathways, as opposed to the unconjugated bile acid pathway described earlier which teams up with LPS to further liver damage and disease. This potent combination greatly increases risk factors for liver disease.

In fact, history bears this out with explorers like John Colter and David Livingston who faced extreme emotional and physical hardship over long periods of time and were documented as dying relatively young from jaundice or liver disease. Current medicine also supports these risk factors in veterans, child abuse survivors, and others who suffer from extended exposure to extreme adverse conditions, making lifestyle intervention a top priority early on.

As we can see, a cascading sequence of liver ailments like fatty liver, fibrosis, cirrhosis, and even cancer along with ensuing pancreas, kidney, and heart issues can not only manifest from poor diet over time, but can actually be initiated by chronic emotional stressors, wherein, a bad diet exacerbates metabolic consequences even more.

Nutrition can play a top role in calming these stress pathways by buffering cortisol and nerve support. Therefore, obtaining adequate amounts of nutrient like phosphatidylserine (PS), Vitamin C, magnesium, B vitamins, and omega-3s to lower cortisol and support parasympathetic nerve health, along with exercise/physical activity and unwinding from stress are key and foundational to begin reversing damage and moving forward. (270. DiagnosTechs, 2016)

Relieving chronic stressors by various modes such as prayer, therapy, decompressing in nature and/or quiet places, grounding techniques, alternative treatments, and better sleep are also known to help with stress, so aside from waking hours, this becomes the next goal, starting with ways to try and improve our sleep.

Sleep – Sleep is God's gift to humanity (and animals) that allows our conscious mind to rest, recharge, and partition a fresh new day away from the concerns of the previous one. It's clear that calming the mind by relieving stress as outlined above can help improve our sleep, but conversely, biohacking our way to better sleep apart from reducing the effects of stressful events can also help to buffer stress.

In today's world, people are searching high and low for ways to improve their sleep from taking supplements to avoiding artificial blue light, listening to calming sounds, and more. By now most of us know that reading blue screens on phones and computers too late can alter circadian rhythms and promote insomnia or at minimum, poorer sleep quality. Although still being proven, many individuals can attest to the positive effects of shielding wireless routers in bedrooms and keeping cell phones several feet away during sleep.

In light of these sleep efforts, it helps to understand the science behind sleep. Sleep is generally defined as an altered conscious state that allows our bodies and brain to heal and rejuvenate, supporting brain function and overall health. Whereas deep sleep supports bodily tissue repair and the immune system, REM and N2 sleep support memory consolidation, emotional processing, and recalibration of the brain synapses. (271. Patel AK, 2025)

The pattern of sleep progresses in various stages from light to deep and back to light again, but unlike a common misconception of one long U-shaped cycle, sleep actually follows many cycles per night.

The various stages consist of light non-REM sleep or N1, followed by intermediate N2, and then deeper N3, before transitioning to REM defined by dreaming and an intermediate state of sleep. This sequence occurs several times a night and is characterized by a rising sinc wave or sawtooth-like pattern that gradually progresses toward lighter sleep and looks something like the graphic figure below.

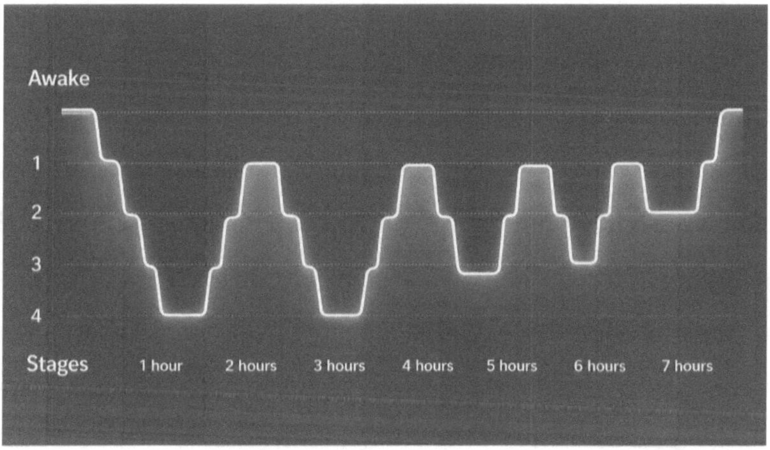

Figure 16. Sleep Cycle

Functional sleep involves several areas of the brain including the hypothalamus for controlling sleep onset, hippocampus for memory consolidation during dreaming, amygdala for processing emotions during dreaming and reducing wakeful stress, thalamus for suppressing sensory signals during sleep, and brainstem for regulating transitions related to REM sleep and wakefulness.

GABA, an important neurotransmitter, is another crucial factor for sleep hygiene that activates receptors in both the hypothalamus and brain stem, promoting sleep by regulating stimulatory neurons.

Accordingly, nutrition plays a big part in quality of sleep, so this can be an important compensating factor for those struggling with sleep issues. For one, GABA can be biosynthesized from glutamate using B_6 as a cofactor, but it can also be obtained from supplements or foods like pumpkin seeds, strawberries, adzuki beans, wheat germ, and barley bran. However, endogenous GABA is ideally sufficient given enough glutamate and B_6 from the diet. (272. Heli Z, 2022)

Everyone has heard of melatonin and for good reason. It's one of the most clinically supported nutrients for promoting better sleep. So, which is more abundant, endogenous melatonin produced from serotonin or melatonin from food and supplements? To start with, melatonin is produced both in the pineal gland of the brain and the gut, but only pineal melatonin is responsible for regulating sleep, whereas gut melatonin is a powerful antioxidant, anti-inflammatory agent, and regulates autophagy and apoptosis. (273. Xie X, 2022)

Ideally, melatonin can be biosynthesized from serotonin utilizing several nutrients and enzymes and again, we can see the role of Mg, B vitamins, and SAM. The chart below shows this basic pathway:

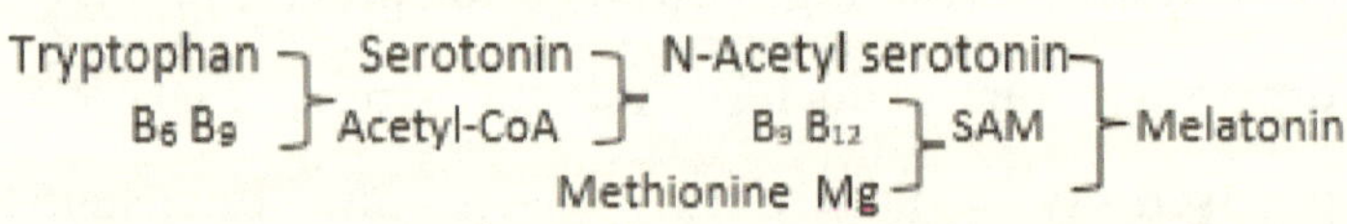

For younger people and those getting adequate tryptophan, magnesium, and B vitamins, the serotonin pathway may be enough, but for others, supplements may be needed including iodine which may help with decalcifying the pineal gland. (274. Group, E, 2017)

Magnesium and glycine are two other important nutrients that support sleep quality. Magnesium helps by not only supporting GABA and melatonin production via SAM, but also by regulating the hypothalamus and calming the nervous system. Glycine works as a neurotransmitter promoting relaxation and sleep onset, but also to regulate body temperature. Together, magnesium and glycine work synergistically as magnesium glycinate to combine benefits.

Other nutrients that may help with sleep in addition to GABA, melatonin, magnesium, and glycine include those in foods containing omega-3s, vitamin D, zinc, beta carotene, and fiber. Nutrient cofactors like B vitamins can be obtained from foods like meats, whole grains, and legumes, but should only be taken as supplements during the day, not night, so as to stimulate the brain.

Specific foods noted for all these nutrients include pistachios, tart cherries, and eggs for melatonin; seafood, poultry, and pork for tryptophan; spinach, avocados, sweet potatoes, and nuts for magnesium; fatty fish and shellfish for vitamin D and zinc; along with foods that calm the gut and support healthy digestion like fermented foods that promote probiotics. (275. Suni, 2025)

Sleep is so key because it comprises one third of our lives, allowing us to rejuvenate and start each day fresh. Along with nutrition, exercise, and biohacking; easing worry and relying on God's provision can help us get the rest we need. "In peace I will lie down and sleep, for you alone, O LORD, will keep me safe." Psalm 4:8

Alternative Health Modes

As complex creatures, the many aspects to maintaining our health are multi-faceted. While nutrition, exercise, sleep and stress reduction are foundational pillars for health, inspiration from God and discoveries for medicinal interventions throughout human history have revealed many modes of what today we call alternative treatments, apart from modern medicine. Some require professional expertise while others are so basic we can do them ourselves.

Included among these are chiropractic, sauna/cold plunge, red light therapy, and grounding along with acupuncture and herbal therapy which were reviewed earlier. Luckily, many of these topics are well covered and familiar to the public, but a basic review is still helpful.

Chiropractic – You either love it or you don't. In the absence of any other spinal manipulation method, chiropractic adjustments are a very effective treatment for not only long term spinal alignment to prevent or slow the degeneration of vertebrae and discs, but also for relieving pinched nerves which can affect all organs including the gut, brain, muscles, and just about every area of the body connected to the central and peripheral nervous systems. (276. Coleman, 2023)

In addition to musculoskeletal health, chiropractic care may also help relieve autonomic nervous system issues like indigestion, respiratory, and even long term cardiovascular effects via nerve decompression; as well as positive modulating benefits to the HPA axis, hormones, and immune system. (277. Kovanur Sampath K, 2015)

Once thought to be an unsubstantiated pseudoscience decades ago, chiropractic care has since earned an evidence-based reputation as an effective treatment mode and integrated into many healthcare systems today. Understandably, many people, including those with serious musculoskeletal conditions are skeptical or afraid of outright physical manipulation of their spines, but thanks to newer methods, less jarring techniques are available, although slightly less effective.

Spring-loaded hand devices called actuators can be used to pop in vertebrae with a gentler approach than manipulation, but may require more treatments. Stretching and massaging beforehand can relax muscle enough to reduce discomfort during adjustments also.

Whatever the technique or comfort level, chiropractic care has been shown to be an effective treatment for spinal as well as overall nerve and organ health when it comes to spinal compression. I personally have been getting it periodically most of my life whether for injuries or for spinal maintenance and can attest to the benefits of preserving strength, mobility, and posture, so it may be something to consider.

Sauna and Cold Plunge – While many holistic insights bridge connections between East and West, one in particular originates from the traditional North. Since ancient times, Nordic cultures like the Finnish have practiced contrast therapy, aka, sauna steam heat combined with immersion in cold water. Sauna was achieved by pouring water over heated stones to produce steam in enclosed huts while cold plunging was done in cold water or by rolling in snow.

Aside from modern sauna structures and cold plunge tanks, not much has changed since then, except that the appeal of these modes has been popularized worldwide. Each mode has their own, albeit similar health benefits, but together they provide a synergistic therapy that promotes cardiovascular health, reduced cholesterol, increased fat burning, and lowering of stress. (278. Sim, n.d.)

Sauna works by raising body temperature, thus increasing heart rate in order to deliver blood flow to the skin surface which can then dissipate heat by sweat triggered from the hypothalamus. This process, which mimics the action of mild to moderate exercise, increases blood circulation, thereby clearing cholesterol from the blood and arterial walls while at the same time, lowering cortisol by reducing stress and promoting relaxation via the parasympathetic nervous system and release of endorphins.

Cold plunging works by shocking the body and producing noradrenaline which breaks down fat to be used for heat and over time, reducing cholesterol as well. The cold also triggers conversion of white fat to brown fat or BAT, which then will burn itself for heat as well. These processes serve to promote fat loss and increase metabolic rates, enhancing energy, focus, and mood.

Together, sauna and cold plunge on a regular basis have been shown to not only reduce cholesterol and CVD risk, body fat, and stress, but also to reduce all-cause mortality and increase longevity. Far more than just spa treatments, these modes are backed by science.

Red Light Therapy – Without a doubt, red light therapy is one of the hottest health topics of this decade. Although it sounds like a fad, there is actually a lot of science behind it. For one, the sun itself emits beneficial red and infrared light (IR), along with all the other wavelengths in the spectrum including visible and ultraviolet light, but today, many red light devices are readily available to the public.

Red light devices come in many shapes and sizes and are specifically designed to emit frequencies that optimize wavelengths to include red light and IR typically between 600 and 950 nanometers (nm). Devices range from saunas and tanning-like beds to home devices such as panels, masks, pads, and lamps.

Red light on the low end of this range from 600-700 nm is used to rejuvenate near surface issues like fine wrinkles, sun-damaged skin, surface wounds, and hair follicles; whereas, IR in the 780-950 nm range is used to treat deeper tissues like muscles, joints, bones, and even cognitive issues, though research is still in the process of codifying these deeper benefits. (279. González-Muñoz A, 2023)

Despite these benefits, health experts suggest that given the right schedule and available sunlight, a person can get similar results by basking in the sun at certain times of the day. Although sunrise and sunset project a much higher ratio of red and IR light than midday, they also project much lower intensity than midday. The trade-off is that UV is also much higher at midday, but some of the concern over higher skin cancer risk may be alleviated by lifestyle changes.

Research shows that excess omega-6 fats in cellular membranes can trigger inflammation and increased risk of skin cancer as shown by not only western diets in the West, but also a higher prevalence in sunbaked regions like the Middle-East where historic risk was low.

Furthermore, C:15 fatty acid and omega-3s are shown to be more protective of the skin, and over a few years of cleaning up the diet, can replace excess omega-6s in cell membranes, making the skin more resilient to UV exposure and promoting lower cancer risk. Nonetheless, omega-3s may not filter out basal cell risk, but do suppress the more dangerous melanoma, whereas C:15 appears to reduce risk of all forms according to research. (280. Park MK, 2018)

Whether by focused red light therapy devices or by consistent, but regulated and informed exposure to the sun that promotes natural vitamin D production in a state of health free of excess omega-6s, red and infrared light are shown to promote healing in many ways.

Rejuvenating skin, soft tissue, and relieving pain with red and IR light as well as supporting circadian rhythm by replacing blue light with red or amber fire light in the evening can reset our internal clock and bring us closer to nature's rhythms the way God intended.

Grounding or Earthing – Most alternative therapies, including the ones just covered may at first appear unsupported, but are actually backed by science, and grounding is no different. Although it sounds king of bohemian, grounding is an electrical potential reality that promotes the flow of electrons from the earth to humans.

Global thunderstorms around the earth create a small electrical charge flow between the ionosphere and earth called the global electric current, which couples with very low EMF waves that circle the earth called Schumann Resonances. This creates a stable global current while local variation of this field, caused by solar activity between sunrise and sunset, helps regulate our circadian rhythms. Studies show that subjects retain these rhythms even when shielded underground from the sun, but while grounded. (281. Koniver, 2023)

As long as people have walked the face of the earth, grounding has occurred. Simply walk barefoot or touch any area of your body to the earth and you are grounding. Today, however, most people are insulated from the ground with shoes, and depending on whether they live on ground level and/or non-earth material flooring, may never actually touch the ground for days, weeks, or even months!

The result can lead to congested inflammation in the body that's hard to dissipate, potentially requiring the body to overcompensate with additional antioxidants, other nutrients, and more stress reduction methods than would otherwise be necessary if grounding is already part of a healthy regime. Tapping into the earth can help.

Research reveals that grounding provides a wide range of benefits like cardiovascular health via better circulation, reduced viscosity, and improved heart rate variability while also boosting the immune system. Moreover, grounding supports brain activity, quality sleep, and stress and sex hormone balancing caused by inflammation.

Reducing inflammation and oxidative stress through grounding also improves mitochondrial function and ATP energy production while reducing cellular damage which can lead to many chronic diseases along with accelerated aging. (282. Giulivi C, 2025)

Since inflammation reduction affects the whole body, grounding has also been shown to help alleviate digestive issues and conditions tied to our nervous system such as anxiety, depression, insomnia, and even back pain. Today, there are many devices are available such as grounding mats, pads, and cuff links for those unable to go outside for weather or medical reasons or who find it impractical, but for those who can, getting outside is ideal.

Walking the dog, time at the beach, swimming in bodies of water, relaxing at the park or in the backyard are some of the many simple ways a person can ground. For those more activity inclined, there are even grounding shoes and boots available to expand the options. Whatever the case, grounding can be a simple, but profound way to tap into the earth's natural rhythm to help support long term health.

In this section, we have covered a whole variety of alternative health modes that can be paired to provide amazing health benefits, in addition to the incredible benefits of acupuncture and herbs covered in Part 4. By integrating both traditional and modern techniques from East, West, and North, we can promote better health and bring ourselves a little bit closer to nature while tapping into the healing power of the earth that God originally intended for us.

Part 6 – Nutritional, Organ, and Health Summaries

All through this study, we've covered a comprehensive spectrum on how various modes of health and nutrition can help change the direction of our health and set the stage for healing in whatever capacity that might look like for us and our particular circumstances, including proper medical care when needed.

In this final topical part of the book, we'll synergize and summarize data from all the covered research, cited references, and other sources into categorized summaries, rankings, and bullet points showing how all the nutrients, foods, and health factors correlate with organs, metabolic ailments, and overall path to better health.

Whereas this first segment details all the benefits and deficiencies for specific nutrients, the second segment builds on this information by summarizing nutritional and holistic health strategies according to various organs, ailments, and desired health outcomes. Quite long, Part 6 reads like a comprehensive two-part reference manual.

The Most Important Foundational Nutrients

Ranked by System-Wide Impact and Potency

Throughout my research, certain nutrients kept re-surfacing again and again, not just as supporters, but as foundational regulators. While all nutrients are essential on some level, these rankings are a best-effort analysis based on all the cumulative research in this book that should help shed light on how to begin assessing health goals.

Fermented Foods and Probiotics – Though not strictly nutrients, we've seen how healthy gut biota stand at the apex of systemic health, assisting the liver in remaining resilient so that as a pair, the gut and liver can efficiently regulate whole-body homeostasis.

With that in mind, this guide describes the most potent nutrients based on system-wide benefits from the most foundational, down to more specialized nutrients. Scarcity or abundance in the diet is not a rating factor, only efficacy is. The ranking considers:

(1) # of organs/systems affected, (2) critical regulatory functions, (3) deficiency consequences, and (4) overall metabolic impact.

Tier 1: THE MASTER REGULATORS

The nutrients with the most profound, system-wide impact

1. MAGNESIUM

The Undisputed King

Why it's #1: Involved in over 300 reactions in the body. Teams up with B1 and B3 as cofactors in a synergistic relationship. Regulates electrolytes, energy production, neurotransmitters, DNA repair, and PTH production which controls Vitamin D metabolism.

What It Affects:

- **Energy Production:** Converts food to ATP; required for mitochondrial function; cofactors with B vitamins

- **Electrolyte Balance:** Works with PTH to regulate calcium, potassium, and sodium
- **Heart Health:** Regulates heart rhythm and blood pressure; supports heart muscle function
- **Nervous System:** Totally essential for neurotransmitter production and autonomic nervous system
- **Bone Health:** Required for calcium homeostasis
- **Muscle Function:** Prevents cramping and supports muscle contraction/relaxation
- **DNA/Protein Synthesis:** Critical for DNA repair, formation, and protein creation
- **Vitamin D & Calcium Activation:** PTH and magnesium together regulate Vitamin D metabolism
- **Blood Sugar & Cholesterol:** Improves insulin sensitivity and cholesterol management
- **Stress & Sleep Regulation:** Calms nervous system; supports HPA axis; prevents adrenal fatigue
- **Anti-Inflammation:** Reduces systemic inflammation
- **Sex Hormones:** Important cofactor for fertility in both men and women
- **B Vitamin Synergy:** Required to convert B1 (thiamine) to its active form (TPP); B1 cannot function without magnesium; upregulates with B3

Deficiency Signs: Muscle cramps, heart palpitations, high blood pressure, anxiety, poor sleep, calcification, kidney stones, fatigue, insomnia, panic attacks

Forms & Benefits:

- Magnesium glycinate: Anxiety, insomnia, sleep
- Magnesium citrate: Digestion (may have laxative effects)
- Magnesium taurate: Blood pressure and blood sugar
- Magnesium L-threonate: Brain support, depression, Alzheimer's
- Magnesium orotate: Athletic performance, energy, cardio support

Sources: Leafy greens, nuts, seeds, whole grains, dark chocolate, avocados, legumes, coconut or mineralized water

Critical Note: Salt deficiency depletes magnesium via increased urination, so maintaining proper sodium/potassium levels is key.

****2. IRON****

The Foundational Builder

Why it's #2: For a nutrient that is routinely added to processed foods, it's no wonder how foundational iron is, though we often take it for granted. Affects DNA synthesis, hormones, muscle growth, bone health, and oxygen transport. Liver regulates iron via hepcidin for systemic whole-body homeostasis, so liver health is paramount.

What It Affects:

- **Red Blood Cell Formation:** Oxygen transport, all tissues
- **DNA Synthesis & Repair:** Vital for cell division, repair
- **Metabolic Homeostasis:** Central regulation via hepcidin
- **Hormone Production:** Required for thyroid and other hormones
- **Muscle Growth:** Essential for muscle protein synthesis and strength
- **Bone Health:** Required for bone formation
- **Brain Function:** Oxygen delivery, neurotransmitter synthesis
- **Heart Health:** Dysregulation can cause heart failure
- **Immune Function:** Critical for immune cell production
- **Energy Levels:** Supports fatigue and cellular respiration
- **Thyroid Function:** Required with iodine and tyrosine for hormone production
- **Pituitary Gland:** Needed with other minerals to function
- **P450 Enzyme Synthesis:** Essential for cytochrome P450 enzyme production, critical for hormone synthesis and detoxification

Deficiency Signs: Anemia, muscle weakness, shallow breathing, restless leg syndrome, sunken eyes/cheeks, brain fog, impaired muscle growth

Overload Issues: Metabolic dysfunction, organ damage (heart, brain, kidneys, lungs) if liver can't regulate properly

Iron Absorption Helpers: Vitamin C, fermented foods, meat, resveratrol (inhibit hepcidin)

Iron Inhibitors: Phytic acid/phytates, polyphenols/tannins, calcium, milk, eggs

Sources: Red meat, organ meats, wild game, fish, poultry; plant sources (spinach, legumes) less bioavailable

Critical Note: Gut health is crucial for regulating iron uptake since no excretion path exists. Copper and Vitamin A regulate iron. Key for those switching to whole food diets who may suffer subclinical deficiency without knowing. Supplements by practitioner approval.

****3. B VITAMINS (Complex)****

The Energy Production & Nervous System Team

Why they're #3: B vitamins are essential and hailed for everything from energy production, heart and nerve health, immune system & red blood cell support, skin, hair, and eye health, brain support, and just overall sense of well-being. Critical cofactors with magnesium.

What They Affect As a Whole:

- **Energy Production:** Convert carbs, fats, proteins to ATP; B3 merges into Krebs cycle in mitochondria

- **Autonomic Nervous System:** B1 critical for sustaining the ANS which affects practically every system in the body
- **Neurotransmitter Synthesis:** Required for serotonin, dopamine, GABA, etc.
- **DNA Synthesis & Repair:** B9 and B12 especially critical
- **Red Blood Cell Formation:** B12 and B9 prevent anemia
- **Nervous System Health:** B12 for myelin sheath; B6 for nerve function
- **Hormone Production:** Required for thyroid, adrenal hormones, sex hormones
- **Homocysteine Regulation:** B6, B9, B12 prevent cardiovascular damage
- **Mood & Mental Health:** Deficiencies linked to depression, anxiety, panic attacks
- **Methylation:** B9 (folate) critical methyl donor
- **Liver Function:** Support detoxification pathways; B1 stored in liver for short time
- **Exercise Recovery:** Depleted by intense exercise; required for muscle recovery
- **Synergy with Magnesium:** B1 and B3 share "special synergistic relationship" - they upregulate each other

B1 (Thiamine) - "Extremely essential nutrient"

- Critical for ANS affecting practically every system

- Magnesium needed to help B1 convert glucose to energy
- Only stored 2-3 weeks in liver; near daily intake necessary
- Deficiency manifests as: enlarged heart, edema, difficult breathing, sleep apnea, neuropathy, slow digestion, brain fog, insomnia, panic attack, fatigue, muscle loss
- Less abundant in common foods than other B vitamins
- **Best sources:** rice bran, spirulina, seeds (flax, sunflower, hemp), nuts (pinion, macadamia, pistachio, pecan), pork, oats, salmon, peas, black beans, brown rice, whole wheat

B3 (Niacin) - "Star of the show"

- Precursor to NAD for mitochondrial energy production
- Works in multiple places in energy cycle and Krebs cycle
- Teams up with magnesium to enable energy process
- Benefits: increased muscle mass, decreased fat, enhanced strength, improved glucose metabolism
- Additional: cholesterol regulation, brain health, skin protection, osteoarthritis relief

B6 (Pyridoxine)

- Neurotransmitter synthesis
- Protein metabolism
- Required for cysteine biosynthesis from methionine
- Cofactor for many processes

B9 (Folate)

- DNA synthesis
- Methylation
- Homocysteine regulation
- Critical during pregnancy

B12 (Cobalamin)

- Nerve function and myelin sheath
- DNA synthesis
- Red blood cells
- Prevents anemia
- Restless leg syndrome prevention

Deficiency Signs:

- General: Fatigue, irritability, anemia, cognitive issues, weakness, neuropathy, depression, restless legs

Sources: Meat, eggs, dairy, fish, leafy greens, legumes, seeds, nuts, whole grains (NOT fortified processed foods - synthetic versions are less effective)

Critical Note: B1 is depleted by metabolization of sugar and processed carbs. Naturally occurring in whole wheat and brown rice (removed in processing). This is why consuming too much processed grains and rice can be detrimental to our health over time.

4. SODIUM & POTASSIUM

The Electrolyte Power Couple

Why they're #4: Vitally important for controlling fluid balance, blood pressure, cellular function, and nerve signaling. The sodium-to-potassium ratio is more important than absolute salt levels.

What They Affect:

- **Fluid Balance:** Regulate water retention and blood volume via sodium reabsorption via RAS system
- **Blood Pressure:** Artery flexibility, vascular tone, and nitric oxide regulation
- **Cellular Function:** Essential for nerve signaling and muscle contraction
- **Kidney Function:** Regulate reabsorption of water via AVP and angiotensin pathways
- **Adrenal Health:** Low salt stresses adrenals by overproducing aldosterone/cortisol
- **Mineral Retention:** Salt deficiency causes wasting of magnesium, potassium from bones/muscles
- **Brain Signaling:** Cross blood-brain barrier freely; essential for neurotransmission
- **Hydration Status:** Regulate osmolality and thirst mechanisms (elderly particularly vulnerable)
- **Blood Sugar Control:** Ample salt boost insulin sensitivity

- **Sleep Quality:** Proper balance supports better sleep
- **Digestive Function:** Support stomach acid and enzyme production
- **Overall Health Benefits:** Better skin conditions, stress hormone reduction

Deficiency Signs (Low Salt): Dehydration, muscle weakness, dizziness, low blood pressure, adrenal stress, mineral wasting, poor sleep, elevated stress hormones, loss of health benefits

Current Research: Minimum 3,000-4,000+ mg sodium daily (1⅓+ tsp salt) for most people without hypertension and depending on activity levels — research shows low salt without hypertension carries same mortality risk as high salt with hypertension

Critical Note: Old paradigm of 1,500-2,300 mg shown far too low. Ratio to potassium matters more than absolute levels. Important for athletes, elderly, people on whole food diets.

Sources:

- Sodium: Unrefined sea salt, Celtic salt, Himalayan salt
- Potassium: Legumes/beans, avocados, dried fruits, leafy greens, potatoes, fish, bananas, dairy, orange juice

5. PHOSPHOLIPIDS, MUFAs, & OMEGA-3 FATTY ACIDS

The Cellular Foundation & Anti-Inflammatory Team

Why they're #5: One of the most crucial building blocks for cells, chemical signaling, mitochondria, and cofactors for sustaining metabolic health. Omega-3 and monounsaturated fats (MUFAs) are the ideal carrier form when integrated into phospholipids (PLs).

What They Affect:

- **Cell Membranes:** Support selective passage of contents in/out of every cell in the body; phospholipids provide structure while omega-3s provide fluidity and anti-inflammatory properties
- **Mitochondrial Function:** Essential for energy production and cellular respiration; PC 40%, PE 30% of mitochondrial membranes
- **Liver Health:** Highest profile of PC in body (aside from lungs); prevents fatty liver
- **Brain Health:** Critical for cognitive function, memory, neurotransmission; DHA makes up myelin sheath; neuroprotection
- **Heart Health:** PC second most abundant in the heart; omega-3s reduce CVD risk and improve heart rhythm
- **Kidney Health:** PC abundant in kidneys
- **Lung Health:** Highest PC content in body; PG also high
- **G.I. Tract:** PE makes up 90% of PL content in gut
- **Chemical Signaling:** Essential for overall cell-to-cell communication

- **Inflammation Control:** Omega-3s powerfully anti-inflammatory; regulate cytokines
- **Eye Health:** DHA concentrated in retina
- **Joint Health:** Omega-3s reduce arthritis and joint inflammation
- **Cellular Senescence:** Omega-3s slow cellular aging
- **Mental Health:** Reduce depression and anxiety
- **Immune Balance:** Modulate immune response
- **mRNA Toxicity Removal:** Recent research shows PC and PE can "encapsulate toxins within cells and remove them" - may help vaccine injury

Key Phospholipid Types:

PC (Phosphatidylcholine):

- Most abundant in liver (apex of whole-body homeostasis), lungs, heart, kidneys, brain
- Primary source of non-dietary choline
- Essential for fat metabolism and preventing fatty liver
- Indirect supply of neurotransmitter synthesis (acetylcholine)
- Works with taurine for bile production
- Only 30% of RDA from biosynthesis (PE + methionine)

PE (Phosphatidylethanolamine):

- 90% of G.I. tract phospholipids - critical for gut health

- First in heart, kidneys, brain (second in liver)
- Can be biosynthesized to PC
- MUFAS-infused PE ideal for brain

PS (Phosphatidylserine):

- Brain health and cognitive function
- Stress response and cortisol regulation
- Memory and learning

PI (Phosphatidylinositol):

- Cell signaling pathways
- Insulin function

PG (Phosphatidylglycerol):

- Precursor to cardiolipin for mitophagy
- Second most abundant in lungs
- PA (phosphatidic acid) from cabbage is a precursor

CL (Cardiolipin):

- Vital for the very mitochondrial cell function itself
- Cell wall integrity
- Energy production
- Mitophagy (mitochondrial autophagy)

Omega-3 Fatty Acids (EPA/DHA):

- "Star of the show" for brain health and inflammation

- "PLs are the ideal carrier form of omega-3s"
- Protects cardiolipin
- Omega-3 infused PC from wild fatty fish, caviar, and eggs
- Essential for cell membranes, especially in brain and heart
- Modern diet has unhealthy ratios of omega-6s vs omega-3s

Sources: May be taken as supplements

- PC: Pastured eggs (highest), soy, liver, beef, chicken, pork, chickpeas, tuna, spinach, seeds/nuts
- PE: Pastured eggs (highest), soy, liver, pork, chicken, tuna, beef, chickpeas, spinach, seeds/nuts
- PG: Soy, oats, spinach, pastured eggs, buckwheat, chickpeas, barley; Precursor: Cabbage (PA source)
- Omega-3 infused PC: Wild fatty fish (salmon, sardines, mackerel), caviar, pastured eggs, algae, walnuts
- Note: Farmed fish contain lower omega-3 content

Monounsaturated Fats:

- Suppress pro-inflammatory cytokines
- Support cardiovascular health
- Support brain health and cognitive function
- Reduce inflammation and improve lung function
- Help regulate calcium homeostasis for kidney function

Sources: Olive oil, avocados, nuts, seeds

Critical Note: At the apex of whole-body homeostasis, liver contains the highest profile of phosphatidylcholine (PC), aside from the lungs. PLs infused with omega-3s and MUFAs are ideal carriers. Recent discovery about mRNA toxicity removal is welcome news to many suffering adverse consequences from the pandemic. Grain-fed animals have unhealthy omega-6:omega-3 ratios.

6. ZINC

The Immune & Growth Master

Why it's #6: Essential mineral required for 300+ enzymes. Critical for immune function, growth hormone, testosterone, wound healing, and iron regulation. One of most foundational minerals for health, vitality, and robust hormones.

What It Affects:

- **Immune Function:** Essential for immune cell production and function, especially T-cell maturation
- **Lung Health:** "May be most potent" nutrient for lung disease prevention
- **Growth Hormone Production:** Critical for human growth hormone (HGH) synthesis; essential for growth, vitality, and anti-aging
- **Sex Hormone Production:** Required for testosterone synthesis; reproductive health and vitality in both sexes
- **Wound Healing:** Critical for tissue repair

- **DNA Synthesis:** Required for cell division and growth
- **Protein Synthesis:** Essential for building muscle, tissue
- **Enzyme Function:** Cofactor for 300+ enzymes
- **Brain Function:** Neurotransmitters; memory; learning
- **Skin Health:** Treats acne; wound healing
- **Gut Health:** Maintains intestinal barrier integrity; required for healthy G.I. tissue (with glutamine, glycine, arginine, Vitamin A)
- **Vision:** Boosts eye health; prevents macular degeneration
- **Taste & Smell:** Required for sensory function
- **Iron Regulation:** Works with copper to regulate iron homeostasis
- **AMPK Pathway:** Mitigates metabolic issues
- **Pituitary and Thyroid Support:** Required with other nutrients for pituitary and thyroid function, T3 production

Deficiency Signs: Frequent infections, slow wound healing, hair loss, skin issues, poor taste/smell, growth delays

Sources: Oysters, red meat, poultry, beans, nuts, whole grains

Critical Note: Exercise depletes zinc. Required in larger amounts for weight lifters. Most potent for lung disease and immune health along with Vitamins A, C, and D.

Tier 2: THE ESSENTIAL REGULATORS

Critically important for major organ systems and metabolic pathways

7. TAURINE

The Multi-System Powerhouse

Why it's #7: One of the most abundant and free amino acids in the body. Essential for bile synthesis, heart health, brain function, and mitochondrial health. Listed among most crucial cofactors for sustaining metabolic health and longevity. Works synergistically with magnesium and choline.

What It Affects:

- **Bile Synthesis:** Conjugates with cholesterol to form bile salts; works with choline
- **Heart Health:** Concentrated in heart tissue; regulates calcium and contractility
- **Brain Health:** Essential neurotransmitter function; neuroprotection; concentrated in brain
- **Eye Health:** Highly concentrated in retina; prevents macular degeneration
- **Mitochondrial Function:** Supports cellular energy
- **Skeletal Muscle:** Improves exercise performance and recovery; effects similar to niacin

- **Blood Sugar:** Enhances insulin sensitivity; improves glucose metabolism
- **Liver Support:** Works with choline for fat metabolism and bile
- **Detoxification:** Bile conjugation for toxin removal
- **Immune Function:** Supports immune response
- **Longevity:** Research shows anti-aging benefits
- **Calcium Regulation:** Modulates cellular calcium levels
- **Muscle Mass & Fat Loss:** Similar to niacin's effects

Deficiency Signs: Poor bile flow, heart problems, vision issues, neurological symptoms, muscle weakness, poor exercise recovery

Sources: Meat, fish, seafood, poultry, eggs, dairy; less in plant foods

Biosynthesis: Can be made from cysteine and methionine (methionine → cysteine → taurine pathway)

Critical Note: Taurine and choline tag-team in the liver but follow several parallel paths. Both provide similar benefits for metabolic homeostasis, muscle support, cellular integrity, and brain health.

8. CHOLINE

The Liver's Best Friend & PL Precursor

Why it's #8: Essential PC metabolite, but also bi-directional source of PC that merits special attention for its foundational role.

Critical in preventing fatty liver for downstream health, the brain, energy, and nervous system; acting as a cofactor in several ways.

What It Affects:

- **Liver Health:** Prevents fatty liver; essential for fat metabolism and export
- **PC Production:** Dietary source of free choline primarily from plant-based foods for phosphatidylcholine synthesis
- **Bile Production:** Required for fat digestion and detoxification; works with taurine
- **Cell Membranes:** PC in mitochondrial membranes for energy production
- **Brain Health:** Neurotransmitter (acetylcholine) synthesis; memory and cognition
- **Mitochondrial Function:** Supports energy production through PC
- **Fat Metabolism:** Prevents fat accumulation in liver
- **Detoxification:** Supports Phase 2 liver detoxification
- **Muscle Support:** Works with taurine for muscle function
- **Heart Health:** Maintains cardiovascular function via vagal activity and reduced inflammation
- **Gene Expression:** Methyl donor via choline metabolism

Deficiency Signs: Fatty liver, poor fat digestion, cognitive decline, nervous system dysfunction, and muscle damage

Sources (free choline): Wheat germ, soy, potatoes, Brussel sprouts, broccoli, cauliflower, collard greens, beef, and poultry

Critical Note: Only 30% of RDA comes from biosynthesis (from PC via PE + methionine). Must get from diet. Teams up with taurine for promoting bile synthesis and liver resilience

9. VITAMIN D

The Immune & Bone Regulator

Why it's #9: Responsible for hundreds of functions in the body and behaves more like a hormone than a vitamin. Critical for immune function, bone, brain and organ health. Synthesized from cholesterol via the skin from sunlight or obtained by diet; both activated by the liver and kidneys. Pivotal in CVD prevention.

What It Affects:

- **Immune System:** Pivotal in immune response; reduces inflammation
- **Cardiovascular Health:** Pivotal in CVD prevention; reduces heart disease risk
- **Bone Health:** Essential for calcium absorption and preventing osteoporosis
- **Calcium Homeostasis:** Works with PTH, magnesium, and potassium to regulate calcium

- **Cholesterol Metabolism:** Sunlight may help reduce extra cholesterol since Vitamin D is synthesized from cholesterol
- **Brain Health:** Neuroprotection, mood regulation, cognitive function
- **Muscle Function:** Required for muscle strength and recovery
- **Kidney Function:** Supports kidney health and mineral balance
- **Metabolic Health:** Improves insulin sensitivity
- **Hormone regulation:** Supports pituitary, sex hormones
- **Lung Health:** Protective against respiratory infections; important for lung support
- **Gut-Liver Connection:** Sunlight indirectly linked to gut health as precursor to serotonin production

Deficiency Signs: Weak bones, frequent infections, fatigue, depression, muscle pain, restless leg syndrome

Sources: Sunlight exposure (primary), fatty fish, cod liver oil, egg yolks, mushrooms

Critical Note: Requires magnesium for activation. PTH and magnesium together regulate Vitamin D metabolism. Cofactor with magnesium for calcium regulation. Sunlight has been directly linked to gut health, circadian rhythm, and nitric oxide production

10. VITAMIN C

The Universal Protector

Why it's #10: One of the few nutrients that crosses the blood-brain barrier freely. Plays such an important role in so many processes. Essential for immune function, collagen synthesis, and iron uptake.

What It Affects:

- **Immune System:** Powerful antioxidant; supports white blood cell function
- **Collagen Production:** Essential for skin, joints, blood vessels, bones
- **Iron Absorption:** Boosts iron uptake, inhibiting hepcidin
- **Antioxidant Role:** Protects cells from oxidative stress
- **Brain Health & Blood-Brain Barrier:** Neuroprotection
- **Adrenal Support:** Required for cortisol and adrenaline
- **Heart Health:** Protects blood vessels; reduces CVD risk
- **Lung Health:** Most potent for lung disease along with zinc and Vitamin D; protects against pneumonia; COPD
- **Wound Healing:** Speeds tissue repair
- **Neurotransmitter Synthesis:** Required for dopamine, norepinephrine synthesis
- **Detoxification:** Supports liver detox pathways; primarily phase 1, but also phase 2, enhancing glutathione's effect

Deficiency Signs: Frequent infections, slow wound healing, bleeding gums, fatigue, joint pain

Sources: Citrus fruits, berries, peppers, broccoli, leafy greens, fermented foods

Critical Note: Works synergistically with fermented foods for iron absorption. Most important iron absorbing foods are vitamin C/fermented foods, meat, and resveratrol.

Tier 3: THE SPECIALIZED ESSENTIALS

Highly important for specific systems or conditionally essential

11. GLYCINE

The Versatile Performer

Why it's #11: Important conditionally essential amino acid and neurotransmitter. Critical for collagen, bile synthesis, glutathione, and creatine production. Listed among most crucial cofactors.

What It Affects:

- **Collagen Production:** One-third of collagen is glycine; joint/skin health, but also essential for structural bone matrix
- **Neurotransmitter Function:** Calms anxiety; reduces depression; improves sleep; co-transmits with glutamate
- **Bile Synthesis:** Combines with taurine to form bile salts

- **Glutathione Production:** Along with cysteine and glutamine - protective of most organs
- **Creatine Production:** For muscle energy and strength
- **Blood Sugar Regulation:** Improves insulin sensitivity
- **Gut Health:** Required for healthy G.I. tissue; supports intestinal barrier
- **Sleep Quality:** Promotes relaxation and better sleep
- **Nervous System Function:** Important amino acid and neurotransmitter for NS
- **Methylation:** Involved in one-carbon metabolism

Deficiency Signs: Poor sleep, joint pain, anxiety, slow healing

Sources: Bone broth, gelatin, collagen, meat with connective tissue

Critical Note: Often deficient in modern diets lacking bone broths and organ meats. Critical for gut health, collagen production, bile synthesis, glutathione, and creatine.

12. GLUTAMINE

The Gut Healer

Why it's #12: Crucial for gut health and immune function. Can be conditionally essential during stress or illness. Required for maintaining healthy G.I. tissue.

What It Affects:

- **Gut Health:** Main fuel for intestinal cells, heals leaky gut
- **Immune Function:** Fuels immune cells, crucial in illness
- **Wound Healing:** Speeds tissue repair
- **Muscle Recovery:** Prevents muscle breakdown
- **Brain Function:** Crosses BBB; precursor to glutamate (neurotransmitter) and GABA
- **Nitrogen Transport:** Moves nitrogen between tissues
- **Acid-Base Balance:** Helps regulate pH
- **α-KG Production:** Intermediate between glutamine and glutamate that regulates anabolic/catabolic balance
- **Liver Support:** Gluconeogenesis and detoxification
- **Glutathione Formation:** Along with cysteine and glycine forms antioxidant
- **Thirst & Fluid Balance:** Precursor/metabolite glutamate signals thirst to brain; GABA quenches thirst

Deficiency Signs: Leaky gut, frequent infections, poor wound healing, muscle wasting

Sources: Meat, fish, eggs, dairy, cabbage, bone broth

Critical Note: Required for maintaining healthy G.I. tissue along with glycine, arginine, zinc, and Vitamin A. Very central amino acid for multiple reactions in both directions.

13. TYROSINE

The Thyroid, Adrenal, & Brain Booster

Why it's #13: Precursor to dopamine, thyroid hormones, and adrenal hormones. Crucial end-product is CoQ10.

What It Affects:

- **Energy Production:** CoQ10 vital for mitochondrial ATP
- **Thyroid Function:** Required for T3 and T4 hormone production (with iodine, iron, copper)
- **Dopamine Production:** Neurotransmitter for focus, motivation, pleasure
- **Adrenal Hormones:** Precursor to adrenaline and noradrenaline
- **Stress Response:** Supports mental performance under stress
- **Focus & Motivation:** Improves cognitive function
- **CoQ10 Organ Support:** Supports heart, kidneys, liver, pancreas, lungs, brain
- **Antioxidant Function:** Via CoQ10 pathway which protects cardiolipin for mitophagy

Deficiency Signs: Low thyroid function, poor focus, low motivation, fatigue, advanced liver disease

Sources: Meat, fish, eggs, dairy, nuts, seeds, soy

Critical Note: High levels of tyrosine can signal advanced liver disease (not cause of it) from impaired liver enzymes that fail to catabolize excess into energy. Liver disease impairs kidneys which are tasked with regulating tyrosine among other things.

****14. METHIONINE & CYSTEINE****

The Sulfur Amino Acid Duo

Why they're #14: Critical pathway to taurine production and methylation. Vital for glutathione, among most crucial cofactors.

What They Affect:

- **Taurine Production:** Methionine → cysteine → taurine
- **Anti-inflammatory:** Boost anti-inflammatory benefits
- **Methylation:** Methionine provides methyl groups
- **Glutathione:** Cysteine is limiting factor for glutathione production (with glycine and glutamine)
- **Neuroprotection:** Brain protective effects
- **Liver Health:** Required for choline synthesis (PC can be made from PE + methionine)
- **Bile Synthesis:** Via taurine pathway
- **Detoxification:** Glutathione supports liver detox

Sources: Eggs, meat, fish, dairy, Brazil nuts

Critical Note: Listed among most crucial cofactors. Methionine pathway to taurine provides proper bile synthesis; mitochondrial, heart, and brain health; blood sugar, fitness performance, immune support, and longevity.

****15. SELENIUM****

The Thyroid Protector

Why it's #15: Essential for thyroid hormone conversion and powerful antioxidant function. Required for CoQ10 biosynthesis.

What It Affects:

- **Thyroid Function:** Converts T4 to active T3 with help from iron and zinc; protects thyroid tissue
- **Antioxidant Defense:** Key component of glutathione peroxidase
- **Immune Function:** Supports immune response
- **CoQ10 Biosynthesis:** Required for production along with tyrosine
- **Heart Health:** Protects against oxidative damage
- **Cancer Prevention:** Antioxidant and immune support
- **Detoxification:** Supports glutathione function
- **Reproductive Health:** Important for fertility
- **Exercise Recovery:** Reduces oxidative stress

Deficiency Signs: Thyroid problems, fatigue, weak immune system, muscle weakness

Sources: Brazil nuts (highest), seafood, organ meats, eggs, sunflower seeds

Critical Note: Narrow therapeutic window - easy to get too much. 1-3 Brazil nuts daily is sufficient. Exercise depletes selenium.

****16. IODINE****

The Thyroid Essential

Why it's #16: Essential component of thyroid hormones. Critical for metabolism regulation.

What It Affects:

- **Thyroid Hormones:** Required for T3 and T4 production
- **Metabolic Rate:** Regulates whole-body metabolism via thyroid
- **Brain Development:** Critical during pregnancy and childhood
- **Energy Levels:** Supports cellular energy production
- **Body Temperature:** Regulates thermoregulation via thyroid
- **Detoxification:** Supports removal of heavy metals
- **Breast Health:** Concentrated in breast tissue

- **Immune Function:** Antimicrobial properties
- **Pituitary-Thyroid Communication:** Low iodine negatively affects communication

Deficiency Signs: Hypothyroidism, fatigue, weight gain, cold sensitivity, goiter

Sources: Seaweed, seafood, iodized salt, eggs, dairy

Critical Note: Low levels of vitamin A and iodine can negatively affect communication of the pituitary gland in order to make TSH. Works with tyrosine, iron, copper, manganese, zinc, Vitamin E.

17. CALCIUM

The Bone & Signaling Mineral

Why it's #17: Essential for bones but also critical for cellular signaling. Must be properly regulated to prevent calcification.

What It Affects:

- **Bone Health:** 99% of body's calcium in bones and teeth
- **Muscle Contraction:** Required for heart and skeletal muscle function
- **Nerve Signaling:** Essential for neurotransmitter release
- **Blood Clotting:** Required for coagulation cascade
- **Cellular Signaling:** Second messenger in many pathways
- **Heart Rhythm:** Regulates cardiac contraction

- **Hormone Secretion:** Triggers hormone release
- **Enzyme Function:** Cofactor for various enzymes

Regulation Critical: Excess unregulated calcium causes calcification of arteries, kidneys, tissues. Must be balanced with magnesium, Vitamin D, potassium, and Vitamin K2.

Deficiency Signs: Weak bones, muscle cramps, osteoporosis (though calcification can occur simultaneously in soft tissues)

Sources: Dairy, leafy greens, fish with bones (sardines), almonds

Critical Note: Adequate levels of calcium tend to be overlooked because of adverse effects from calcification experienced by many today. Unless prescribed by a doctor, better to obtain through a whole foods diet rather than supplements or processed foods. Proper regulation from key nutrients above prevents calcification.

18. COPPER

The Iron Regulator

Why it's #18: Regulates iron in the body. Essential for connective tissue, energy production, and nervous system.

What It Affects:

- **Iron Regulation:** Critical for iron homeostasis; copper also regulates iron in the body, so iron deficiency, if not from consumption, can result from copper deficiency

- **Connective Tissue:** Required for collagen and elastin
- **Energy Production:** Component of cytochrome c oxidase in mitochondria
- **Nervous System:** Required for myelin formation
- **Thyroid Function:** Required for T4 synthesis
- **Antioxidant Function:** Component of SOD which protects spinal disc health
- **Immune Function:** Supports white blood cell production
- **Heart Health:** Maintains blood vessel integrity
- **Pigmentation:** Required for melanin production
- **Brain Function:** Neurotransmitter synthesis
- **Liver Storage:** Primarily stored in liver and kidneys;

- **Balance Critical:** Blocked bile ducts or gallbladder can prevent excretion of copper leading to copper toxicity from the liver, iron deficiency, and other symptoms in the rest of the body. Excess goes through kidneys via AMPK pathway.

Deficiency Signs: Iron deficiency (despite iron intake), sunken eyes, vision issues, weak connective tissue, premature gray hair

Sources: Organ meats, shellfish, nuts, seeds, dark chocolate

Critical Note: AMPK (triggered by copper excess) is essential enzyme for triggering autophagy and provides alternate pathway for liver/bile regeneration.

19. RESVERATROL

The System-Wide Disease Fighter

Why it's #19: Powerful polyphenol with amazing system-wide benefits on the body and has been shown to relieve conditions in just about every organ in the body. Listed alongside omega-3s and P450 enzyme nutrients as among the most crucial cofactors for sustaining metabolic health.

What It Affects:

- **Heart Health:** Well-known for cardiovascular benefits; reduces CVD risk
- **Kidney Health:** Supports kidney health and function
- **Lung Health:** Great for lung health; protective effects
- **Brain Health:** "Great for brain health"; neuroprotection
- **Liver Disease:** Mitigates liver disease and fibrosis
- **Metabolic Syndrome:** Basically covers most of the metabolically downstream organs
- **Apoptosis Promotion:** Triggers programmed cell death in damaged cells
- **Insulin Sensitivity:** Improves glucose metabolism
- **Fat Metabolism:** Supports weight loss
- **Longevity:** Activates sirtuins for anti-aging
- **AMPK Pathway:** Activates autophagy and cellular regeneration

- **Antioxidant:** Along with quercetin from red wine
- **Anti-inflammatory:** Systemic inflammation reduction
- **Iron Absorption:** Boosts iron uptake, inhibiting hepcidin
- **Gut-Microbiome Modulation:** Regulates intestinal microbiota to improve cognitive function; creates gut-kidney-brain axis connection
- **Intestinal Barrier Protection:** Protects intestinal barrier and prevents intestinal inflammation

Sources: Itadori, red grapes/wine, berries, nuts, dark chocolate

Synergy: Red wine also contains quercetin which provides added antioxidant and anti-inflammatory properties.

Critical Note: It's fantastic for heart, kidney, lung, and brain health which basically covers most of the metabolically downstream organs, but it doesn't stop there. Ancient wisdom of diluting wine (3:1 water/wine) balances hydration with polyphenol benefits. **Therapeutic results require supplements.**

Tier 4: THE SUPPORTING CAST

Important nutrients with more targeted or specialized functions

20. CoQ10

The Mitochondrial Energizer

What It Affects:

- **Energy Production:** Essential coenzyme for ATP production in mitochondria, especially the heart
- **Antioxidant:** Reduces oxidative stress and inflammation
- **Multi-Organ Support:** Supports heart, kidney, liver, pancreas, lung, & brain health
- **Blood Pressure:** Helps regulate vascular function
- **Exercise Performance:** Improves energy and recovery

Biosynthesis: End-product of tyrosine pathway (requires tyrosine and selenium), healthy liver promotes maximum biosynthesis

Sources: Organ meats, fatty fish, whole grains, supplements

Critical Note: Tyrosine pathway shows how end-product CoQ10 supports heart, kidney, liver, pancreas, lung, & brain health along with reducing inflammation and oxidative stress.

21. Vitamin A

The Vision & Immunity Vitamin

What It Affects:

- **Vision:** Essential for retinal function
- **Immune Function:** Supports immune cell production
- **Gut Health:** Required for maintaining healthy G.I. tissue (with glutamine, glycine, arginine, zinc)

- **Skin Health:** Cell growth and repair
- **Reproductive Health:** Critical for development
- **Lung Health:** Important for lung support
- **Thyroid Function:** Low levels negatively affect pituitary-thyroid communication

Sources: Liver, egg yolk, butter, sweet potatoes, carrots, leafy greens

Critical Note: Low levels of vitamin A and iodine can negatively affect communication between your pituitary gland and thyroid. Vitamin A also helps regulate iron. Required for RE recovery.

22. Vitamin E

The Fat-Soluble Antioxidant

What It Affects:

- **Antioxidant Protection:** Protects cell membranes from oxidative damage
- **Immune Function:** Supports immune response
- **Skin Health:** Wound healing and protection
- **Heart Health:** Protects against cardiovascular damage
- **Thyroid Support:** Required with other minerals for thyroid function
- **Lung Health:** Important for lung support

Sources: Wheat germ oil, sunflower seeds, almonds, hazel nuts, avocados, spinach

Critical Note: Required for RE recovery along with Vitamin A and other nutrients.

****23. Vitamin K2****

The Calcium Director

What It Affects:

- **Calcium Regulation:** Directs calcium to bones, away from soft tissues
- **Bone Health:** Activates proteins that bind calcium to bones
- **Cardiovascular Health:** Prevents arterial calcification
- **Dental Health:** Supports tooth mineralization

Sources: Fermented foods (natto highest), grass-fed dairy, egg yolks, organ meats

Critical Note: Essential for preventing calcification when paired with magnesium, Vitamin D, and potassium.

****24. Arginine & Citrulline****

The Nitric Oxide Boosters

What They Affect:

- **Nitric Oxide Production:** Improves blood flow and vascular health

- **Heart Health:** Reduces blood pressure and improves circulation
- **Immune Function:** Supports immune response
- **Fitness Training:** Enhances endurance and recovery
- **Wound Healing:** Promotes tissue repair (arginine)
- **HGH Stimulation:** Upregulates human growth hormone (with lysine and GABA)
- **Kidney Function:** Glycine, ornithine, and arginine biosynthesized in kidneys
- **Gut Health:** Arginine required for healthy G.I. tissue

Citrulline: Precursor to arginine; more effective for supplementation (bypasses liver metabolism)

Sources: Watermelon (citrulline), meat, nuts, seeds, legumes

Critical Note: Arginine works with glutamine, glycine, zinc, and Vitamin A for G.I. tissue health.

25. Alanine

The Carnosine Precursor

What It Affects:

- **Carnosine Production:** Alanine + histidine → carnosine
- **Muscle Buffering:** Carnosine buffers lactic acid during exercise

- **Antioxidant:** Carnosine has antioxidant properties
- **Anti-Glycation:** Prevents protein glycation and AGEs
- **Neuroprotection:** Brain protective effects
- **Central Role:** Very central AA responsible for multiple reactions; bi-directional synthesis with glutamate

Sources: Meat, poultry, fish

Critical Note: Glutamate and alanine are very central AAs responsible for multiple reactions in both directions.

There are so many other nutrients that could be listed – BCAAs like leucine, isoleucine, and valine for muscle protein; minerals like phosphorus and chromium for cells, energy, and metabolism; metabolites like creatine and glutathione for energy, muscles, and antioxidants – the list goes on and on. But with any luck, this guide has given you a more targeted understanding of which powerful nutrients deserve your attention and how they can work together to restore and sustain whole-body health.

In summary, nutrients are absolutely vital and essential for not only physical survival, but also for long-term vibrant health, systemic vigor, and keeping disease at bay. While I believe there are natural solutions by God on this planet for just about every condition, I also feel we should use every tool available to us including modern medical treatment when conditions are advanced or critical, which leads into the second segment of Part 6, recovering from ailments.

Health Strategies for Organs, Ailments, and Systemic Health

Nutritional, Lifestyle, and Integrative Approaches

While the first segment was ranked by nutrients, this segment is listed from the most systemic diseases down to those affecting the foundational organs at the apex of metabolic health. These are end-stage manifestations of chronic metabolic dysfunction, representing years of accumulated damage across multiple organ systems.

What begins as asymptomatic liver issues or leaky gut can progress over decades into cardiovascular disease, diabetes, kidney disease, neurodegenerative disorders, cancer, and the like which is why **Fermented Foods & Probiotics** should be at the apex of every diet.

Addressing these conditions requires a comprehensive approach that targets nutritional support at every level – not just symptoms – but the root metabolic imbalances involving the gut-liver-brain axis, chronic inflammation, oxidative stress, and hormonal disruption in order to help reverse metabolic cascades and restore optimal health.

MOST SYSTEMIC & SERIOUS METABOLIC DISEASES

Cardiovascular Disease and Disorders

Heart disease remains the #1 killer and represents one of the final stages of metabolic breakdown. It develops via multiple pathways: plaque in the arteries, damaged blood vessels from inflammation, calcium deposits from mineral imbalances, and disrupted cholesterol metabolism from unhealthy bile and nutrient deficiency.

The key to prevention and reversal lies in supporting healthy bile acid and cholesterol formation, proper nutrition, converting harmful white fat to beneficial brown fat, appropriate exercise, reducing inflammation, and protecting arterial walls.

Key Nutrients:

- **Phospholipids & Omega-3s:** Essential for heart tissue, mitochondria, and arterial walls. Omega-3s reduce triglycerides and inflammation; phospholipids maintain cell membranes.

  Best sources: Pastured eggs, soy, liver, fatty fish, grass-fed meat, sunflower seeds; fatty fish , caviar, algae, walnuts; supplements

- **Taurine, Glycine, Magnesium, Sodium & Potassium:** Support cholesterol metabolism via bile acids; stabilize heart rhythm; regulate blood pressure; maintain electrolyte balance.

  Best sources: Yellowfin tuna, scallops, white fish, dark poultry, for taurine; Bone broth, collagen powder, gelatin for glycine; Leafy greens, nuts, seeds, dark chocolate, legumes, avocados, coconut or mineral water for magnesium; sea salt for sodium; avocados, leafy greens, sweet potatoes, legumes for potassium

- **CoQ10 & B Vitamins:** CoQ10 powers mitochondrial ATP in heart muscle. B6, B9, B12 regulate homocysteine – elevated levels significantly increase CVD risk.

  Best sources: Organ meats, fatty fish, grass-fed meat, pastured eggs, leafy greens, legumes, supplements

- **Vitamins C & K2:** Vitamin C builds collagen that keeps arterial walls strong and flexible. K2 directs calcium to bones and away from arteries, preventing calcification.

  Best sources: Citrus, berries, bell peppers, leafy greens, broccoli; Natto (highest K2), fermented foods, grass-fed dairy, eggs

- **Zinc & Polyphenols:** Zinc reduces atherosclerosis and supports blood vessel integrity. Polyphenols reduce inflammation and oxidative stress that damage arterial walls.

  Best sources: Oysters, beef, pumpkin seeds, dark chocolate, liver, wheat germ; Spices, dark chocolate, berries, flaxseed, olive oil, nuts, red grapes/wine, fruit, whole grains for polyphenols

Overlapping foods: Leafy greens, pastured eggs, fish, legumes, avocados, nuts, seeds, polyphenols, grass-fed meat, organ meats

Therapeutic Supplements:

- **CoQ10:** Therapeutic doses protect heart, critical if on statins
- **Berberine:** Broad protection across many CV health conditions

Support Strategies:

Exercise: Essential. Promotes beneficial brown fat, improves circulation, fights inflammation. HIIT and circuit training work best fit depending on condition, but low intensity is better and safer for those trying to recover from CVD or increase stamina slowly .

Intermittent Fasting & Stress Control: Time-restricted eating activates AMPK for autophagy and reduces inflammation. High cortisol from chronic stress damages blood vessels. Daily prayer/meditation, adaptogenic herbs, and ample sleep are essential.

Cold Exposure: Brief cold showers or ice baths convert white fat to beneficial brown fat for better triglyceride metabolism.

Diabetes & Metabolic Syndrome

Type 2 diabetes is a severe metabolic disruption affecting nearly every organ. It builds over years from ultra-processed foods, sugar, refined carbs, chronic inflammation, gut problems, and fatty liver.

High blood sugar impairs tissue, while the pancreas works overtime producing insulin to try and compensate. Eventually it exhausts. Fatty liver throws off blood sugar control, bile production, and detox, creating a cascade that hits your pancreas, kidneys, heart, and brain. Reversing it means restoring insulin sensitivity, supporting the pancreas, and healing the gut-liver connection.

Key Nutrients:

- **Magnesium, Taurine & Glycine:** Low magnesium impairs insulin sensitivity. Taurine helps regulate blood sugar and glycine improves insulin response.

  Best sources: Leafy greens, nuts, seeds, dark chocolate (85%+), avocados; Seafood, grass-fed meat, bone broth, collagen

- **Alpha-Lipoic Acid & Zinc:** ALA improves insulin sensitivity and prevents diabetic nerve damage. Zinc supports insulin production plus immune function and wound healing

  Best sources: Chia & flax, walnuts for ALA; Oysters, wheat germ, liver, beef, pumpkin seeds, dark chocolate

- **B Vitamins:** B1 is crucial for sugar metabolism, along with B3 which also helps with cholesterol and mitochondrial energy.

  Best sources: rice bran, spirulina, seeds, nuts, pork, oats, salmon, peas, black beans, brown rice, whole wheat

- **Polyphenols:** These powerful compounds reduce inflammation, improve insulin sensitivity, and support healthy blood sugar through several paths to protect against diabetic complications.

  Best sources: Spices, dark chocolate, berries, flaxseed, olive oil, nuts, red grapes/wine, other fruits, whole grains

- **Fiber & Omega-3s:** Soluble fiber slows down sugar absorption, reducing dangerous blood sugar spikes. It also feeds good gut bacteria that along with omega-3s reduces inflammation.

  Best sources: Leafy greens, cruciferous veggies, berries, legumes, resistant starch (cooled potatoes/rice) for fiber; Wild fatty fish, caviar, algae, pastured eggs, walnuts for omega-3s

Overlapping foods: Leafy greens, legumes, grass-fed meat, pastured eggs, seeds/nuts, whole grains, seafood, polyphenols

Therapeutic Supplements:

- **Berberine & Chromium:** Berberine works like metformin for insulin sensitivity, lowers blood sugar, supports gut biome and AMPK. Chromium helps insulin function properly.

  Best sources: Berberine supplements; Chromium supplements or nutritional yeast, broccoli, whole grains, and grass-fed meat

- **GABA & B6:** GABA restores insulin-secreting beta cells and protects against beta-cell death. B6 needed as cofactor.

  Best sources: GABA supplements (lower in fermented foods)

- **Ceylon cinnamon:** Specific for insulin-sensitizing compounds.

Support Strategies:

Eliminate Processed Foods & Sugar: Most important step, non-negotiable. Ultra-processed foods, added sugars, refined flour, and seed oils are the root cause. Cut them out and switch to whole foods.

Intermittent Fasting: Powerfully restores insulin sensitivity, promotes cellular cleanup, reduces liver fat, activates fat-burning. Work with your doctor to advise and adjust medications as needed.

Exercise & Sleep: Both resistance and HIIT dramatically improve insulin sensitivity – muscle contraction pulls glucose out of the blood, independent of insulin resistance. Getting a full night's sleep improves glucose tolerance and managing stress reduces chronic cortisol and blood sugar.

Cancer – Prevention & Support

As the second leading cause of mortality, cancer represents a fundamental breakdown in cellular regulation where damaged cells evade apoptosis and autophagy and won't die off. This metabolic environment plays out through multiple pathways:

- Chronic inflammation damages DNA and feeds tumor growth.
- When cellular cleanup (autophagy) fails, damaged cells persist.
- Unregulated mTOR growth drives uncontrolled multiplication.
- A disrupted gut biome creates immune dysfunction.
- Poor liver detox affects how well you clear carcinogens.
- Mitochondrial problems shift cells toward a sugar-based metabolism that cancer thrives on.
- The focus: reduce inflammation, activate cellular cleanup, support detox, and strengthen immunity.

Important: These are for prevention and complementary support alongside conventional medical treatment. Never replace medical cancer treatment. Some nutrients may interact with specific therapies – always work with your oncology team and practitioner.

Key Support:

- **Curcumin, Sulforaphane & Resveratrol:** Powerful anti-inflammatory and anti-cancer compounds. Curcumin blocks tumor growth, sulforaphane activates detox enzymes, and resveratrol activates cellular cleanup and AMPK pathways.

Best sources: Turmeric with black pepper; Broccoli sprouts (30-50x more sulforaphane than broccoli), Brussels sprouts, cabbage for sulforaphane; Itadori (Japanese knotweed), red grapes/wine (diluted), berries, dark chocolate for resveratrol

- **Green Tea & Omega-3s:** Green tea polyphenols trigger cancer cell death and block tumor blood vessel formation. Omega-3s fight inflammation driving cancer and may inhibit tumors.

  Best sources: Green tea, matcha; Wild fatty fish, caviar, algae, pastured eggs, walnuts for omega-3s

- **Vitamin D, Selenium & Zinc:** Critical for immune function and proper cell regulation – deficiency linked to many cancers. Selenium and zinc power antioxidant-immune defense systems.

  Best sources: Sunlight (15-30 min daily), fatty fish, pastured eggs, supplements; Brazil nuts (1-2 daily), seafood, organ meats; Oysters, wheat germ, liver, beef, pumpkin seeds, dark chocolate

- **Glutathione & Probiotics:** Master antioxidant for detox. Healthy gut bacteria are crucial for first line immune response.

  Best sources: NAC, whey protein, pastured-eggs, liver, cruciferous veggies, garlic, turmeric, onions, wild fatty fish, spinach, asparagus, avocados

Overlapping foods: Wild fatty fish, pastured eggs, cruciferous veggies, polyphenol foods, nuts, dark chocolate

Therapeutic Supplements:

- **Curcumin, Sulforaphane & Resveratrol:** Powerful anti-inflammatory and anti-cancer compounds. Curcumin blocks tumor growth, sulforaphane activates detox enzymes, and resveratrol activates cellular cleanup pathways.

  Best sources: Curcumin supplements with black pepper (turmeric contains some); sulforaphane extract (broccoli sprouts are highest food source); resveratrol supplements (red grapes contain some)

- **Green tea extract (EGCG):** Concentrated polyphenols for cancer prevention; supplements (or 3-4 cups green tea daily)
- **NAC (N-Acetylcysteine):** Boosts glutathione for detoxification and cancer cell apoptosis. Best sources: NAC supplements
- **Medicinal mushroom extracts:** Turkey tail has strong clinical support for immune modulation and anti-tumor effects; Reishi, shiitake, maitake provide general immune support

Support Strategies:

Intermittent fasting: Powerfully activates autophagy or cellular cleanup, reduces inflammation, and may inhibit cancer cells.

Exercise: Reduces cancer risk through multiple AMPK pathways.

Environmental toxins: Weakened immunity, avoid when possible.

Sleep & Stress Reduction: Getting a full night's sleep timed with circadian rhythm is key for immune function and DNA repair. Chronic stress promotes inflammation and suppresses immunity. Managing it through daily prayer/meditation, time in nature, and addressing root causes are crucial.

Neurodegenerative Disease and Nerve Damage

Neurodegenerative diseases represent the brain's version of metabolic dysfunction, developing over decades through multiple interconnected pathways. Brain inflammation from chronic gut issues allows inflammatory compounds past the blood-brain barrier as impaired glucose metabolism starves neurons of energy and damaged proteins accumulate when autophagy fails to cleanup cells.

As a result, mitochondria can't produce ample energy and oxidative stress damages neurons, while critical deficiencies in phospholipids and omega-3s impair brain membranes and myelin nerve sheaths. The gut-liver-brain axis is key – bile acids from liver damage can leak into the hypothalamus, disrupting both the BBB and HPA axis. The peripheral nervous system itself can be damaged from chronic stress and take years to heal, so a patient perspective is key.

Key Nutrients:

- **Phospholipids (PC, PE, PS, PG) & Choline:** Essential for brain cell membranes and myelin insulation around nerves. Choline is the precursor for acetylcholine, a critical neurotransmitter for memory, learning, and attention.

Best sources: PC & PE: Pastured eggs, soy, liver, beef, chicken, pork, chickpeas, tuna; PS: Faty fish, soy, beef, chicken, pastured eggs, PG: Soy, oats, spinach, pastured eggs, buckwheat, chickpeas, barley; PG precursor: Cabbage (PA); supplements; Choline: wheat germ, soy, potatoes, cruciferous veggies

- **Omega-3 (DHA):** Highly concentrated in the brain, integrated into phospholipids for myelin nerve insulation, reducing brain inflammation. Essential for cognitive and memory function.

  Best sources: Fatty fish, caviar, algae, pastured eggs, walnuts

- **Vitamins B & C:** B1, B6, B9, B12 support neurotransmitters, homocysteine levels, and brain energy – buffering brain fog and cognitive issues. Vitamin C protects against oxidative damage.

  Best sources: Grass-fed meat, pastured eggs, fish, leafy greens; citrus, berries, bell peppers, broccoli

- **CoQ10, Curcumin & Resveratrol:** CoQ10 powers energy-hungry neurons – Parkinson's linked to mitochondrial issues. Curcumin crosses into brain, reduces inflammation, and protein dysregulation. Resveratrol protects neurons.

  Best sources: Organ meats, fatty fish, supplements; turmeric with black pepper; red grapes/wine, berries, supplements

Overlapping foods: Fatty fish, pastured eggs, grass-fed meat, liver, chicken, leafy greens (spinach), berries

Therapeutic Supplements:

- **Magnesium L-Threonate:** Magnesium L-threonate crosses the BBB most efficiently, supporting memory and cognition
- **CoQ10, Curcumin (with black pepper) & Resveratrol:** Therapeutic doses for neuroprotection
- **Lion's mane mushroom extract:** Concentrated for nerve growth factor production and neuroplasticity

Support Strategies:

Intermittent fasting is one of the most powerful tools – it activates cleanup of damaged proteins, reduces brain and nerve inflammation, promotes brain growth factors.

A full night's sleep critical – the brain detoxifies during deep sleep. Exercise increases brain growth factors and new cell formation, helping to learn new skills, read, and do puzzles. Maintain social connections as loneliness accelerates decline. Healing the gut also reduces brain inflammation.

Lungs & Respiratory Ailments

The lungs face constant exposure to pollutants, toxins, pathogens, and oxidative stress from breathing roughly 20,000 a day. Lung tissue requires robust antioxidant defenses, healthy mucus membranes, strong immunity, and efficient cellular repair processes.

Poor kidney and metabolic dysfunction compound respiratory problems – fluid retention stresses the lungs, while insulin resistance and chronic inflammation worsen asthma and COPD. Chronic oxidative damage and nutrient deficiencies contribute to conditions like asthma, COPD, pulmonary fibrosis, and lung cancer.

Key Nutrients:

- **Phospholipids (PC, PG), MUFASs, & Omega-3s:** PC and PG are crucial for pulmonary surfactant, preventing lung collapse and allowing normal gas exchange. PG, a cardiolipin precursor, is vital for mitochondrial resilience in lungs. Monounsaturated fats reduce inflammation, improve lung function. Omega-3s reduce inflammation, asthma, and COPD; buffer oxidation.

- **Best sources:** PC: Pastured eggs, soy, liver, beef, chicken, pork, chickpeas, tuna; PG: Soy, oats, spinach, pastured eggs, buckwheat, chickpeas, barley; PG precursor (PA): Cabbage; MUFAs: Olive oil, avocados, nuts; Omega-3s: fatty fish, caviar, algae, pastured eggs, walnuts

- **Glutathione/Vitamin C:** Glutathione is the lungs' master antioxidant for detoxification. Vitamin C is highly concentrated in lung fluid and provides frontline defense against infections and environmental damage.

- **Best sources:** NAC, whey protein, pastured-eggs, liver, cruciferous veggies, garlic, turmeric, onions, fatty fish, spinach, asparagus; citrus, berries, bell peppers, kiwi

- **Zinc & Selenium:** Zinc is critical for respiratory immune defense and helps protects lung tissue. Selenium powers glutathione for antioxidant protection.

  Best sources: Oysters, wheat germ, liver, beef, pumpkin seeds, dark chocolate; Brazil nuts (1-2 daily), seafood

- **Vitamin A & Carotenoids:** Vitamin A helps maintain lung membranes and decrease infection risk. Lycopene protects against oxidative damage from air pollutants and smoking.

  Best sources: Liver, pastured eggs, grass-fed dairy, cod liver oil; sweet potatoes, carrots, pumpkin, leafy greens, tomatoes

- **Magnesium & Vitamin E:** Magnesium relaxes bronchial smooth muscle, reducing airway constriction. Vitamin E protects lung membranes from oxidative damage.

  Best sources: Leafy greens, pumpkin seeds, almonds, dark chocolate, avocados; Wheat germ oil, sunflower seeds, almonds, hazel nuts, avocados, spinach

- **Vitamin D & Quercetin:** Vitamin D boosts immunity in the lungs, reducing infection risk and severity. Quercetin is a powerful antihistamine that stabilizes mast cells, reducing allergic airway responses and asthma symptoms.

  Best sources: Sunlight, fatty fish, pastured eggs, cod liver oil; red onions, apples, berries, capers, green tea

Overlapping foods - Pastured eggs, fatty fish, organ meats, leafy greens (including spinach), cruciferous vegetables (including cabbage), berries, pumpkin seeds, soy, onions.

Therapeutic Supplements:

- **NAC (N-Acetylcysteine):** Precursor for lung glutathione; reduces inflammation and thins mucus.

- **Zinc (for pulmonary fibrosis):** Can arrest or reverse pulmonary fibrosis by inhibiting collagen scar tissue formation.

- **Vitamin D3 & Quercetin:** Vitamin D3 if inadequate sun exposure; quercetin for antihistamine and stabilizing asthma

- **Bromelain & Herb Support:** Mucus-inflammation reducing enzyme (supplement or pineapple core); ginger, turmeric, garlic, oregano for bronchodilator, antimicrobial, and inflammation.

Support Strategies:

Breathing Exercises & Exercise: Diaphragm/controlled breathing strengthens lung muscles and improves oxygen/CO_2 balance; swimming, singing, or light EE exercise for lung capacity

Hydration & Mucus Management: Adequate hydration, steam inhalation with eucalyptus oil, saline rinses (distilled/filtered water)

Air Quality & Other actions: Reduce exposure to mold, dust, pollution, smoking. Consider elimination diet if chronic congestion; manage blood sugar and kidney health.

Bone, Muscle, and Connective Tissue Impairment

Bone and muscle health are intimately connected – muscle contractions stimulate bone formation, while bone provides structural support and mineral reserves for muscle function. Along with cartilage & ligaments that connect them, they require constant restoration, adequate protein synthesis, and proper mineral balance.

Nutrient deficiencies, chronic inflammation, insulin resistance, and hormone imbalances accelerate bone loss (osteoporosis, osteopenia) and muscle wasting (sarcopenia). Poor kidney function disrupts calcium-phosphorus balance and vitamin D activation, further impairing both systems. Sedentary lifestyle, aging, and metabolic dysfunction compound deterioration in both tissues simultaneously.

Key Nutrients:

- **Vitamin D3 & Vitamin K2:** Vitamin D3 essential for calcium absorption and bone mineralization, while also supporting muscle strength and function; K2 directs calcium away from soft tissues and arteries to bones where it belongs.

  Best sources: Sunlight, fatty fish, pastured eggs, cod liver oil, supplements; natto (highest K2), fermented foods, grass-fed dairy (especially aged cheese), pastured eggs

- **Calcium, Magnesium & Phosphorus:** Primary bone minerals working in balance – calcium for structure, magnesium for integration, phosphorus for strength.

Best sources: Sardines, dairy, leafy greens, sesame seeds; leafy greens, pumpkin seeds, almonds, dark chocolate, avocados; fish, meat, pastured eggs, dairy, legumes

- **Protein & Collagen:** Ample protein crucial for muscle and bone matrix support – 0.45 g/lb. (1.0 g/kg) for young and healthy older adults with lower activity to 0.73 g/lb. (1.6 g/kg) for younger athletes and older adults who are frail adults or who perform resistance training. Collagen provides structural matrix for bone mineralization and supports connective tissue for muscles and joints.

  Best sources: Meat, fish, pastured eggs, dairy, legumes; bone broth, collagen powder, gelatin, organ meats

- **Boron & Silicon:** Improves vitamin D, calcium, and magnesium activity, supporting bone density. Silicon for connective tissue and collagen production in bones and joints.

  Best sources: Avocados, almonds, prunes, raisins, dates; oats, barley, leafy greens, green beans, bananas

- **Vitamin C & Zinc:** Vitamin C is essential for collagen synthesis Zinc supports bone formation, muscle protein synthesis, and wound healing while regulating bone remodeling. Both support cartilage and ligament repair and strengthening.

  Best sources: Citrus, berries, bell peppers, broccoli, kiwi; oysters, wheat germ, liver, beef, pumpkin seeds

- **Omega-3s & Vitamin E:** Reduces autoimmune inflammation that drives bone loss and muscle breakdown. Vitamin E protects muscle cells from oxidative damage during exercise and aging.

  Best sources: Fatty fish, caviar, algae, pasture-eggs, walnuts; wheat germ oil, sunflower seeds, almonds, hazel nuts, avocados, spinach

Overlapping foods: - Eggs, fatty fish, leafy greens, avocados, berries, pumpkin seeds, almonds, legumes, grass-fed meat.

Therapeutic Supplements:

- **Vitamin D3 (with K2):** If inadequate sun exposure; critical for calcium absorption and bone mineralization
- **Collagen powder:** If not consuming bone broth regularly
- **Boron & Silicon:** Supplements – Difficult to get therapeutic amounts from food; one source for silicon is horsetail extract

Support Strategies:

Resistance Training: Weight-bearing stress from resistance training a few times a week directly stimulates bone formation and muscle growth. Both tissues respond – "use it or lose it"

Adequate Protein Intake: Eat protein every meal (25-45g per meal) to optimize muscle protein synthesis. Post-exercise protein within 2 hours; bone broth or collagen between meals to optimize.

Optimize Hormone Balance: Testosterone, estrogen, growth hormone, and IGF-1 all support both bone and muscle. Strength training, adequate sleep, stress management, and proper nutrition support healthy hormones naturally. Avoid endocrine disruptors

Reduce Inflammation & Acid Load: Chronic inflammation and high dietary acid – processed foods like vegetable oils, refined flours, high fructose corn syrup, and food chemicals, caffeine, alcohol, and smoking – aid bone and muscle loss. Alkaline foods protect tissues.

Sun Exposure & Outdoor Activity: Natural sunlight provides vitamin D during outdoor weight-bearing activities for bone-strengthening. Walk, hike, or do sports several times weekly.

INTERMEDIATE METABOLIC ORGAN AILMENTS

These organ systems play essential regulatory roles for long-term health, but have somewhat more resilience than Tier 1 organs. Dysfunction here cascades downstream and compounds metabolic issues over time. Supporting these organs prevents systemic deterioration and protects the more critical systems above them.

Immune System: Spleen, Lymphatic System & Thymus

The immune system is your body's natural defense network. The spleen filters blood from systemic circulation, destroys pathogens, and produces white blood cells and antibodies. It also lightens the liver's detoxification load, but gets overwhelmed when the liver fails.

The lymphatic system includes bone marrow (blood cells), thymus (trains T-cells), and lymph nodes (filter threats). The thymus is critical but shrinks with age, explaining increased vulnerability to disease in older adults, but immunol therapy appears promising.

Impaired immunity shows up as frequent infections, autoimmune diseases (lupus, rheumatoid arthritis, MS), chronic inflammation, enlarged spleen, weakened cancer detection, and immune decline. The root is often a damaged gut-liver axis (~80% of immune tissue in gut) and an unhealthy liver disrupts everything else downstream.

Key Nutrients:

- **Vitamin D & Vitamin A:** Vitamin D modulates immune response and enhances T-cell activation; Vitamin A maintains mucosal barriers and supports thymus function

  Best sources: Sunlight, fatty fish, cod liver oil, pastured eggs; liver, pastured eggs, butter, carrots, sweet potatoes

- **B6, B9, B12, SAMe & Choline:** Power methylation cycle to dial down genetic immune overreaction driving autoimmune disease, while retraining immune cells against self-attack

  Best sources: Organ meats, grass-fed meat and dairy, fish, pastured eggs, leafy greens, legumes, sunflower seeds

- **Zinc & Selenium**: Zinc vital to train T-cells and prevent thymus atrophy; selenium protects immune cells and against cancer

Best sources: Oysters, wheat germ, liver, beef, pumpkin seeds, dark chocolate; Brazil nuts (1-2 day), seafood, pastured eggs

- **Omega-3s & Polyphenols**: Reduces chronic inflammation which suppresses immunity; supports lymphocyte health and regulates inflammation

  Best sources: Wild fatty fish, caviar, algae, pastured eggs, walnuts; berries, green tea, dark chocolate, grapes, olive oil

- **Fiber:** Feeds good gut bacteria that produce SCFAs (butyrate) supporting immune function by preventing leaky gut

  Best sources: Vegetables, legumes, whole grains, nuts, seeds, fruit for fiber; (butter, fermented dairy for direct butyrate)

- **Glutathione**: Protects immune cells during pathogen response

  Best sources: Whey protein, pastured-eggs, liver, cruciferous veggies, garlic, turmeric, onions, fatty fish, spinach, asparagus

- **Vitamin C & Vitamin E:** Vitamin C supports white blood cells. Vitamin E protects immune cells and T-cell function.

  Best sources: Citrus, berries, bell peppers, leafy greens, broccoli; nuts, seeds, avocados, olive oil

Overlapping Foods: Wild fatty fish, pastured eggs, liver, spinach/green vegetables, berries, grass-fed meat, shellfish, nuts, seeds, citrus, dark chocolate, olive oil, fermented foods, avocados

Therapeutic Supplements:

- **Zinc & Selenium:** Zinc for T-cell development; selenium for glutathione
- **NAC:** Master antioxidant for immune protection.
- **Medicinal mushroom extracts:** Turkey tail has strong clinical support for immune modulation and thymus support; Reishi for general immune support.

Support Strategies:

Gut, Liver, & Immunity: Prioritize gut health (~80% of immune tissue in gut – probiotics, fermented foods heal leaky gut), support the liver (impaired liver stresses spleen), ample protein for immune cell rebuilding, avoid immune suppressors (excess alcohol, smoking, needless antibiotics), support adrenals for rapid immune response

Stress, Exercise, & Sleep: Manage stress (cortisol suppresses immunity), moderate exercise (enhances lymphatic flow – avoid overtraining), full night's sleep (immune repair during deep sleep).

Kidney Disease

While the liver filters blood from digestive organs (and processes systemic blood also), the kidneys filter the entire blood supply including that pre-filtered through the liver first. As mighty filtration plants, they filter blood continuously, removing waste while regulating electrolytes, blood pressure, and fluid balance.

They also activate vitamin D for calcium absorption and immune function, while producing hormones like erythropoietin (EPO), that regulate red blood cell production in bone marrow.

When kidney function declines, consequences cascade – weakened bones and immune system, unbalanced electrolytes and retained fluid that burdens the lungs and heart, and toxins that damage the brain and liver. Supporting liver, gut, adrenals, and thyroid is key. CKD is hard to reverse once advanced – early intervention is critical.

Key Nutrients:

- **Resveratrol:** Powerful polyphenol with amazing system-wide restoration benefits, but especially for protecting the kidneys, without the balancing act required of other nutrients for CKD

- **Best sources:** Itadori, red grapes/wine (diluted 3:1), berries, nuts, dark chocolate; but disease likely require supplements

- **Omega-3s & Phospholipids (PE/PC):** - PE and PC are main kidney membrane constituents; omega-3s reduce inflammation, protect against diabetic nephropathy

- **Best sources:** Wild fatty fish, caviar, algae, pastured eggs, walnuts; pastured eggs, soy, liver, grass-fed meat, chicken

- **Magnesium & Potassium:** - Critical for electrolyte balance, protects from oxidative stress, prevents stones; regulates BP – potassium balanced by magnesium, but moderate in CKD

- **Best sources:** Leafy greens, nuts, seeds, whole grains, dark chocolate, avocados, legumes, coconut or mineralized water; legumes, avocados, dried fruits, leafy greens, root vegetables, fish, bananas, dairy, orange juice

- **Vitamin D & B Vitamins:** - Kidneys activate calcitriol (active Vitamin D); crucial for kidneys; B6/B12/folate regulate high homocysteine (damages blood vessels)

- **Best sources:** Sunlight, fatty fish, cod liver oil, pastured eggs; organ meats, grass-fed meat, fish, leafy greens

- **Vitamin C, E & Selenium:** - Vitamin C provides frontline antioxidant protection against oxidative damage; Vitamin E protects membranes; selenium powers glutathione

- **Best sources:** Citrus, berries, leafy greens, bell peppers; nuts, seeds, avocados; Wheat germ oil, sunflower seeds, almonds, hazel nuts, avocados, spinach; Brazil nuts, seafood

- **MUFAs & CoQ10:** - Olive oil, avocados, nuts for healthy fats; CoQ10 supports mitochondrial health

- **Best sources:** Olive oil, avocados, nuts; organ meats, fatty fish,

Overlapping Foods: Fatty fish, pastured eggs, organ meats, leafy greens (spinach), berries, avocados, nuts, citrus, dark chocolate, red grapes

Therapeutic Supplements:

- **TUDCA:** Powerful bile acid for mitigating CKD
- **Resveratrol:** Only supplements contain therapeutic amounts

Support Strategies:

Diet and Exercise: Balanced electrolytes, intermittent fasting, a Mediterranean or keto-light diet, and exercise address upstream inflammation, improves blood flow, and muscle-mediated insulin

Pancreatic Ailments

The pancreas serves dual functions, yet is largely a digestive organ. The small endocrine part produces insulin and glucagon (covered in Diabetes section), but the larger part produces digestive enzymes: lipase for fats, protease for proteins, and amylase for carbohydrates.

Poor enzyme production leads to malabsorption, deficiencies, and loss of fat-soluble vitamins. Research shows a 70% correlation with NAFLD and fatty pancreas (NAFPD), which affects enzymes. Chronic inflammation can damage pancreatic cells irreversibly.

Key Nutrients (Broad and Exocrine-Focused):

- **Zinc** - Critical for enzyme production; prevents autodigestion inside the pancreas; protects cells from oxidative damage; prevents scar tissue formation. Pancreatic disease depletes zinc.
- **Best sources:** Oysters, wheat germ, liver, beef, pumpkin seeds, dark chocolate

- **Cysteine & Glutathione Support:** Master antioxidant, critical for protecting pancreatic tissue from oxidative stress

  Best sources: Fatty fish, meat, dairy, whey protein, pastured-eggs, liver, cruciferous veggies, garlic, turmeric, onions, spinach

- **Cruciferous Flavonoids & Probiotics:** - Sulforaphane reduces inflammatory oxidative stress; probiotics aid pancreas via the gut

  Best sources: Broccoli, cabbage, kale, cauliflower; Brussels sprouts, fermented foods (sauerkraut, kimchi, kefir, yogurt)

- **Polyphenols:** - Broad support for oxidative stress and enzymes

  Best sources: Spices, dark chocolate, berries, flaxseed, olive oil, nuts, red grapes/wine, other fruits, whole grains

- **Vitamin C & Vitamin K:** - Essential antioxidant protection, support enzyme production, reduce inflammation

  Best sources: Citrus, berries, bell peppers, broccoli, leafy greens; leafy greens, fermented foods, Brussels sprouts

Overlapping Foods: Leafy greens, berries, broccoli, Brussels sprouts, citrus, olive oil, fermented foods, organ meats, pastured eggs, poultry, fish

Therapeutic Supplements:

- **Berberine:** Can powerfully mitigate pancreatitis, fatty pancreas, fibrosis, and cancer via AMPK activation and apoptosis. Can reduce gut microbes, but needs to be cycled to avoid gut issues.

- **NAC**: Forms glutathione; shields pancreas from oxidative stress
- **Digestive enzymes:** For true pancreatic enzyme insufficiency, but long term use for other ailments may reduce biosynthesis

Support Strategies:

Multiple: Addresses NAFLD, avoid alcohol and gallstone triggers, intermittent fasting for pancreatic rest, support vagus nerve (nutrients, deep breathing, cold plunge), shun processed food/sugar

Thyroid Dysfunction

The thyroid makes T4 and T3 hormones that regulate metabolism, energy production, body temperature, and mitochondrial function, but it also impacts the heart, brain, kidneys, gut, and sex hormones.

Hypothyroidism (or iron deficiency) causes fatigue, weight gain, cold intolerance, brain fog, and depression. Hyperthyroidism causes anxiety, weight loss, rapid heartbeat, and muscle weakness. Intestinal permeability driven by chronic stress and poor diet is the most common upstream cause of autoimmune thyroid disease.

Key Nutrients:

- **Vitamin D & Vitamin A** – A and D linked to autoimmune thyroid conditions via gut health; vitamin A and D strengthen the gut barrier and calm immune reactivity at its source

  Best sources: Sunlight, fatty fish, cod liver oil, pastured eggs; liver, pastured eggs, butter, carrots, sweet potatoes

- **Iodine & Selenium:** Iodine is central for thyroid hormones (deficiency causes hypothyroid; excess causes autoimmune); selenium converts T4 to T3 and protects thyroid tissue

  Best sources: Seaweed (kelp, nori), fish, pastured eggs, iodized salt; Brazil nuts (1-3 daily), seafood, organ meats, pastured eggs, sunflower seeds

- **Omega-3s & Vitamin E** - Curbs autoimmune inflammation; Vitamin E protects thyroid tissue from oxidative damage

  Best sources: Wild fatty fish, caviar, algae, pastured eggs, walnuts; wheat germ oil, sunflower seeds, almonds, hazel nuts, avocados, spinach

- **Copper, Zinc & Iron:** - Support T4 to T3 conversion and thyroid function, but zinc also vital for tight junctions in gut to protect against autoimmunity; lack of iron mimics hypothyroid

  Best sources: Liver, oysters, sesame seeds, dark chocolate; oysters, wheat germ, liver, beef, pumpkin seeds, dark chocolate; clams, liver, oysters, seeds, dark chocolate, lentils

- **Tyrosine, B Vitamins, and Choline** - Tyrosine is an amino acid backbone for thyroid hormones; B2/B6/B9/B12 and choline for hormone synthesis, T4-T3 conversion, and methylation

  Best sources: Grass-fed meat & dairy, fish, pastured eggs, nuts, seeds, soy; liver, meat & dairy, fish, pastured eggs, leafy greens

Overlapping Foods: Fatty fish, pastured eggs, liver, grass-fed meat & dairy, oysters/shellfish, Brazil/walnuts, sunflower seeds, pumpkin seeds, leafy greens (spinach), avocados, olive oil, dark chocolate

Therapeutic Supplements:

- **Iodine:** Supplements if deficient (test first) or seaweed
- **Ashwagandha:** Adaptogen for thyroid and autoimmune support

Support Strategies:

Target autoimmune triggers: Swap out refined gluten for whole grains and sourdough, allowing gradual gut re-adaptation; manage stress and exercise (cortisol impairs T4 to T3), avoid fluoride

Gut support: Consume fermented foods, fiber, and probiotics

Adrenal-HPA Dysregulation

The adrenal glands sit atop the kidneys producing cortisol, aldosterone, DHEA, and adrenaline. The HPA axis is disrupted when liver damage allows bile acids to leak into the hypothalamus, damaging the BBB which affects pituitary and adrenal output.

Chronic high cortisol from stress and disruption impairs liver, gut, thyroid, and pancreas causing insulin resistance, belly fat, immune issues, and bone loss. Low cortisol causes fatigue, low blood pressure, salt cravings, and poor stress resilience. Adrenals regulate kidneys via aldosterone; impaired kidneys affect adrenal hormones.

Key Nutrients:

- **Vitamin C & Vitamin B5:** C highly concentrated in adrenals – essential for cortisol synthesis. B5 supports stress hormones.

  Best sources: Citrus, berries, bell peppers, broccoli, leafy greens; Liver, pastured eggs, avocados, mushrooms, sunflower seeds

- **Magnesium & B Vitamins** - Magnesium supports healthy cortisol response and relaxes nervous system; B vitamins are depleted by chronic stress.

  Best sources: Leafy greens, nuts, seeds, dark chocolate, avocados, legumes; Liver, grass-fed meat, fish, pastured eggs

- **Sodium & Potassium** - Aldosterone controls Na-K balance; low aldosterone causes salt cravings and low blood pressure

  Best sources: Unrefined Sea salt (3,000-4,000+ mg sodium); avocados, leafy greens, sweet potatoes, legumes, bananas

- **Omega-3s & Phospholipids** - Reduce stress-induced inflammation, support healthy cortisol response; PC and PS support cell membranes; PS reduces excessive cortisol

  Best sources: Wild fatty fish, caviar, algae, pastured eggs, walnuts; pastured eggs, soy, liver, grass-fed meat, chicken; PS supplements (or sunflower lecithin)

- **Vitamin D & Zinc** - Modulate immune function and inflammation; both depleted by chronic stress

Best sources: Sunlight, fatty fish, cod liver oil, pastured eggs; oysters, wheat germ, liver, beef, pumpkin seeds, dark chocolate

Overlapping Foods: Fatty fish, pastured eggs, organ meats, leafy greens, berries, avocados, nuts, seeds, citrus, broccoli, meat, legumes, dark chocolate, sunflower seeds

Therapeutic Supplements:

- **Phosphatidylserine (PS):** Reduces excessive cortisol
- **Adaptogenic herbs:** Ashwagandha, rhodiola, holy basil, licorice root normalizes cortisol patterns (high or low); cycle herbs to prevent adaptation.

  Best sources: Herbal supplements or teas (cycle herbs)

Support Strategies:

Practice stress management: Prayer, deep breathing, time in nature, therapy, journaling, and adequate sleep supports healthy cortisol rhythm – high in morning, low in evening.

Moderate exercise for adrenal fatigue: Walking, yoga, tai chi, and light endurance exercise support circulation and promote relaxation.

Blood sugar stability and stimulants: Eat protein with each meal, but avoid long fasting if cortisol dysregulated. Avoid caffeine, address chronic inflammation, use adaptogenic herbs as needed, and prioritize rest and recovery.

FOUNDATIONAL ORGANS – The Gut-Liver-Brain Axis

As shown time and again, the liver and gut are foundational for metabolic health. When they're healthy, they protect everything downstream – the thyroid, adrenals, pancreas, immune system, kidneys, heart, brain, lungs, muscles, and bones. When they fail, consequences cascade through every system, including their own.

What often begins as silent liver issues or leaky gut can progress over decades into CVD, diabetes, kidney and lung disease, neurodegenerative disorders, and cancer. Real healing requires addressing the root: the gut-liver-brain axis.

The gut and liver communicate constantly through biochemical signals and nerve pathways. Unhealthy bile from an impaired liver damages the gut lining and alters the microbiome. In turn, a leaky gut sends toxins and inflammatory signals back to the liver, creating a vicious cycle. The vagus and splanchnic nerves connect the gut to the brain, so chronic stress and dysregulation disrupt gut function.

Gut & Microbiome – The Root

Your gut is where nutrition begins and most metabolic disease takes root. It handles digestion and absorption, but equally important: it houses 70-80% of your immune system and produces 95% of your body's serotonin – not your brain. Trillions of gut bacteria influence everything from mood to metabolism, producing vitamins, fatty acids, anti-inflammatory compounds, and neurotransmitters.

Gut dysfunction manifests as leaky gut, Crohn's disease, ulcerative colitis, irritable bowel syndrome (IBS), small intestinal bacterial overgrowth (SIBO), celiac disease, and gastritis. They all share common roots: damaged gut lining, disrupted microbiome, chronic inflammation, and reduced nutrient absorption. Left alone, gut disease drives autoimmunity, food sensitivities, mood disorders, and systemic inflammation that accelerates aging and metabolic decline.

Key Nutrients:

- **Glutamine & Glycine** - Glutamine is the primary fuel for intestinal cells, crucial for healing leaky gut by protecting gut barrier; glycine for collagen (gut lining, connective tissue)

  Best sources: Grass-fed meat, fish, pastured eggs, grass-fed dairy; bone broth, collagen powder, gelatin

- **Phospholipids (PE, PC) & Omega-3s** - PE is 90% of phospholipid content in the G.I. tract; essential for cellular membranes; omega-3s reduce inflammation

  Best sources: Pastured eggs, soy, liver, grass-fed meat, chicken; wild fatty fish, caviar, algae, pastured eggs, walnuts

- **Probiotics & Prebiotics** - Modulate gut biota-bile acid feedback cycle; produce SCFAs (butyrate), vitamins, neurotransmitters; influence bile acids and liver health

  Best sources: Fermented foods (sauerkraut, kimchi, kefir, yogurt), mushrooms, fiber-rich vegetables, diverse diet

- **Vitamin U & Sulforaphane** - Vitamin U from cabbage is remarkable for healing gastritis and ulcers by protecting and repairing gut lining; sulforaphane activates detox pathways, reduces gut inflammation, protects against H. pylori (major cause of gastritis/ulcers), supports gut barrier function

  Best sources: Fresh cabbage juice (highest U – traditional gastritis/ulcer remedy), raw/lightly cooked cabbage, sauerkraut; broccoli sprouts (30-50x more sulforaphane), broccoli, Brussels sprouts, cauliflower, kale

- **Vitamin A & Zinc** – Vitamin A crucial for mucosal and epithelial cell growth, regulates iron uptake; zinc supports gut barrier integrity and immune function

  Best sources: Liver, pastured eggs, butter, carrots, sweet potatoes; oysters, wheat germ, liver, beef, pumpkin seeds, dark chocolate

- **B Vitamins (B2, B6, B12, Folate) & Vitamin C** - Support serotonin production (tryptophan + B6/B12/folate); overall intestinal homeostasis

  Best sources: Organ meats, grass-fed meat, fish, pastured eggs, leafy greens; citrus, berries, bell peppers

- **Vitamin D & Magnesium** - D increases gut microbiota diversity, supports immune function; magnesium supports intestinal motility (movement) and enzyme function

Best sources: Sunlight, wild fatty fish, pastured eggs; leafy greens, nuts, seeds, dark chocolate

Overlapping Foods: Pastured eggs, liver, fatty fish, leafy greens, fermented foods, grass-fed, meat, nuts, seeds

Therapeutic Supplements:

- **Glutamine:** Therapeutic doses necessary for severe gut issues
- **Probiotics:** Specific strains in larger doses needed for dysbiosis
- **Bromelain:** Anti-inflammatory enzyme that protects gut lining and aids digestion. Supplements or pineapple (especially core)
- **MCT Oil (full spectrum):** Reduces microbes and methane gas

Critical Steps:

Eliminate Gut Irritants: Remove processed foods, vegetable oils, added sugars, excess alcohol, unnecessary antibiotics. Consider elimination diet to identify food sensitivities (gluten, dairy).

Support Liver & Manage Stress: Healthy bile is critical – can't heal gut without addressing the liver. Chronic stress damages gut lining and alters microbiome through vagus nerve. Daily stress management (meditation, breathing exercises, therapy when needed) is foundational.

Lifestyle Basics: Chew food well, avoid eating 3-4 hours before bed, ample hydration, a full night's sleep, intermittent fasting once gut is healed, bone broth for gut repair.

The Liver – Master Regulator (& Gallbladder)

As we've seen many times over, the liver is king for metabolic health – the most powerful, resilient, and multifaceted organ, responsible for total metabolic balance, able to fully regenerate itself from just 25% of its original tissue – and for good reason.

Responsible for over 500 essential functions, the liver operates with relative autonomy from the brain, acting as the master regulator that keeps every downstream system in balance. When the liver thrives, everything flows: clean blood, healthy bile, balanced hormones, efficient detoxification, stable blood sugar, and vibrant energy. However, when it fails, the dominoes begin to fall as fatty liver left unchecked can lead to fibrosis, cirrhosis, and other diseases.

Recap of Liver's Critical Roles:

- Blood filtration and processing - Receives blood from all the digestive organs (stomach, intestines, pancreas, gallbladder) and spleen; filters everything before it reaches the rest of the body; pre-filters for the kidneys

- Detoxification powerhouse - Degrades toxins through Phase I and Phase II pathways into harmless water-soluble particles for elimination

- Bile production and recycling - Produces bile for fat digestion and detoxification; recycles 95% through the gut; supplies gallbladder for storage and release

- Hormone regulation hub - Produces or regulates hepcidin (iron), angiotensinogen (blood pressure), T3 (thyroid), aldosterone (kidneys), and adrenal hormones
- Protein and amino acid factory - Biosynthesizes most non-essential and conditionally-essential amino acids; produces plasma proteins (albumin, transferrin, clotting factors)
- Metabolic control center - Processes and stores cholesterol (hormone building block), regulates blood sugar via glycogen, stores vitamins (A, D, E, K, B1, B12, folate) and minerals (iron, copper)
- Master regulator of homeostasis - Orchestrates digestion, detoxification, hormone production, nutrient storage, and metabolic balance for the entire body

The liver receives two blood supplies: the portal vein (consolidates blood from digestive organs and spleen) and the hepatic artery (fresh oxygenated blood from the heart). Remarkably, the liver receives 20-24% of the body's total oxygen – more than any other organ including the brain.

The Bile-Gut-Liver Connection:

Bile production and recycling is central to both liver and gut health. About 95% of bile is recycled back to the liver through the gut where they are reabsorbed, then combined with taurine and glycine to form bile salts before finishing the process with PC to complete bile.

These nutrients are critical – they activate receptors that regulate bile production while cleansing fat and toxins from the liver, so deficiencies contribute to bile dysfunction and fatty liver.

When the liver is impaired, it releases inflammatory signals that damage the kidneys, increases detox burden on kidneys and spleen, secretes unhealthy bile that harms the gut microbiome and intestinal lining (causing leaky gut and malabsorption), and disrupts the brain's (HPA) stress response through bile acids leaking into systemic circulation.

Key Nutrients:

- **Phosphatidylcholine (PC) & Choline**: PC extracted by bile for cholesterol homeostasis; reduces liver fat, regulates triglycerides; heals the liver

  Best sources: Liver, pastured eggs, wheat germ, soybean, pork, beef, fish, chicken, cauliflower, broccoli, Brussels sprouts, potatoes, whole grains, dairy.

- **Methionine & Cysteine**: Precursors to taurine; cysteine critical for glutathione (master antioxidant); anti-inflammatory and neuroprotective

  Best sources: Pastured eggs, poultry, pork, fish, sunflower seeds; red peppers, garlic, onions, broccoli, Brussels sprouts

- **Taurine & Glycine**: Taurine critical for bile salt formation, mitochondrial health, heart, brain, blood sugar, fitness, immune support, longevity; glycine for bile salts, collagen, detoxification

  Best sources: Yellowfin tuna, scallops, mussels, whitefish, dark poultry, red/green algae seaweed; bone broth, collagen, gelatin, grass-fed meat

- **Polyphenols & Resveratrol**: Resolves inflammation, toxins, and fibrosis; resveratrol activates AMPK for autophagy in liver cells, but therapeutic doses require supplements

  Best sources: Spices, dark chocolate, berries, flaxseed, olive oil, nuts, red grapes/wine, other fruits, whole grains; Itadori supplements/tea, red grapes/wine (diluted 3:1), berries, pistachios, dark chocolate

- **B Vitamins (especially B1, B5, B12, Folate)**: B1 needed daily (liver only stores 2-3 weeks); B5 for fat metabolism and detoxification; B12 and folate for DNA synthesis and repair

  Best sources: Organ meats, grass-fed meat, fish, pastured eggs, nutritional yeast, whole grains (B1), leafy greens

- **Magnesium & Zinc**: Crucial for DNA synthesis and repair (with B12, folate, iron); enzyme cofactors

  Best sources: Leafy greens, nuts, seeds, dark chocolate, avocados, legumes; oysters, wheat germ, liver, beef, pumpkin seeds, dark chocolate

- **Vitamins A, D, E, K**: Liver stores; A and D for immune function; E antioxidant protection; K for blood clotting

  Best sources: Liver, pastured eggs, butter, carrots; sunlight, fatty fish, cod liver oil; nuts, seeds, avocados; leafy greens, fermented foods, vitamin D supplement for serious ailments

Overlapping Foods: Eggs, liver, fatty fish, leafy greens, berries, red grapes, grass-fed meat, shellfish, nuts, seeds, dark chocolate, fermented foods, broth, avocados, olive oil

Therapeutic Supplements:

- **TUDCA**: Powerful bile acid for healing liver and gallbladder disease; inhibits liver cell damage and fibrosis that drives fatty liver, supports bile flow, detoxification, and insulin sensitivity

  Best sources: Supplements (synthetically produced); ox bile is another alternative supplement as well

- **NAC**: Forms glutathione to shield liver from oxidative stress and support detoxification; the most clinically validated liver-protective supplement for acute non-metabolic liver damage

- **Berberine:** Protects liver function and reduces metabolic stress on the liver

- **Milk Thistle (Silymarin):** Restores damaged liver cells, reduces inflammation, fibrosis, toxin damage, and chronic liver disease

Support Strategies:

Eliminate Liver Toxins: Remove processed foods, vegetable oils, high fructose corn syrup, excess alcohol. These are direct liver toxins at the root cause of fatty liver epidemic. Switch to clean whole foods.

Intermittent Fasting & Exercise: A powerful interventions for fatty liver. Time-restricted eating activates autophagy and promotes fat burning. Exercise is essential – both resistance training and HIIT.

Sleep & Stress Management: Get ample sleep for liver regeneration – most active repair occurs 1-3 am during deep sleep. In both TCM and current science, suppressed anger and trauma is linked to liver damage. Decompression, prayer, and therapy central to healing.

Support Gallbladder: Gallbladder health mirrors liver health – it stores and concentrates bile. Use all liver strategies above. TUDCA supplements help. Eat moderate healthy fats each meal to stimulate bile release – avoiding fat causes stagnation and more stones.

Summary Conclusion – Largely, metabolic disease and decline result from years of bad dietary habits, stress, lack of exercise and poor sleep. Restorative foods like liver (moderation), eggs, nuts, seeds, leafy greens, fatty fish, cruciferous vegetables, berries, and avocados are central. When we eliminate obstacles to healing and provide essential raw materials for repair, we create conditions that allow the body's innate wisdom to restore balance and vitality. The body has remarkable healing capacity when we properly support it.

Part 7 – Conclusionary Roundup – Tying It All Together

A wise king once said "Wisdom is more precious than rubies, and nothing you desire can compare with her" — Proverbs 8:11. Coined around three thousand years ago, this divinely inspired sage advice can help us thrive and prosper, despite the age we live in with all its societal, environmental, and life style stressors we have to face.

As we begin to wrap up this course, we'll circle back once more to synthesize and summarize all the information we've learned from practical health tips to profound scientific mechanisms supported by ancient wisdom and everything in between. The human body is truly the most complex biological machine in the universe and all of the supporting nutrients, health modes, and natural environment have been exquisitely designed to support it as a dwelling for the essence of who we are and what we are to accomplish here on earth.

My Story – continued

Throughout my quest for answers and insight, this journey has been as much of an unfolding experience for me as for you, the reader. When I woke that dire morning, everything abruptly changed — with stalled digestion, central sleep apnea (spent B1), air hunger, brain fog, and micro-depressions hours after meals, I felt terminal and wondered if I'd ever see the sun shine through the clouds again.

I became emaciated to the point of alarming family and friends, and although by faith and dietary rigor I was beginning to heal, my symptoms and weight didn't bottom out until months after the fact.

My research, coupled with AI forensic analysis of personal history, acute symptoms, and long road to recovery recently revealed that far from just gallstones, fatty liver, and fatty pancreas (as if not enough), I was actually suffering from stage 3 liver fibrosis and beginning of cirrhosis with pancreatic fibrosis at the door step! This was more than a serious ill; it was a cascading metabolic failure in the making that even my positive glass-half-full perspective couldn't shake.

Shortly after the initial crisis is when I began my frenzied quest to overturn every informational rock in order to regain my health; something Big Medicine seemed unwilling or unable to do. This was my mission, my calling — to find natural healing by whatever means it took, and then to pay it forward in order to help educate others the way God would have me do.

Hopefully, I've succeeded in this effort so far, but the journey is not quite over yet, so before heading toward that finish line, let's continue by first exploring a few fun facts and intriguing correlations gleaned from everything we've learned and more.

Did You Know?

Did you know? — That your bones are 30% collagen by volume? Everyone obsesses over calcium, but without the collagen matrix providing structural framework, you can't build strong bones no matter how much calcium you take. This is why bone broth simmered 24+ hours — rich in glycine, proline, and collagen — has been a traditional food for bone health across cultures for millennia.

Did you know? — Pistachios pack more nutrition per bite than almost any food on earth? They're the richest natural source of melatonin, a complete protein with all essential amino acids, highest nut in potassium, loaded with lutein and zeaxanthin for eye health, plus vitamins B6, E, and K. The BCAAs and melatonin in pistachios support muscles and sleep. Ancient Israelites had access to them (Genesis), and they remain a nutritional powerhouse today.

Did you know? — Popeye was right! Spinach isn't just high in iron; it's one of the most metabolically complete foods around. Rich in magnesium (300+ reactions), B9 (methylation), vitamins K and C, lutein (eye health), nitric oxide-boosting nitrates for heart health, antioxidants, and PG for longevity. Popeye's biceps may have been cartoon fiction, but his food choice (cooked) was nutritional genius.

Did you know? — Seven foods dominate disease prevention across the organ ailments in Part 6? **Liver, eggs, nuts & seeds, leafy greens, fatty fish, cruciferous vegetables, and berries** appear repeatedly in CVD, diabetes, kidney disease, cancer, and neurodegeneration – boosting AMPK pathways autophagy, fat burning, and longevity.

Did you know? — Cold exposure after exercise creates powerful synergistic effects? Exercise — especially HIIT or moderate aerobic — followed immediately with cold exposure amplifies brown fat conversion, boosting post-exercise calorie burn by 15-25%., along with improved insulin sensitivity. (Caveat: Delay cold exposure 2-4 hours after resistance training to preserve anabolic signaling.)

Did you know? — Five nutrients can shut down the stress-disease cascade at its source? Chronic stress → cortisol excess → damaged gut barrier → bacterial LPS floods liver → systemic inflammation cascades into autoimmunity, diabetes, heart disease, kidney disease, depression, and neurodegeneration. Rather than chasing dozens of downstream symptoms, five foundational nutrients close the gate:

Magnesium (calms HPA axis, powers 300+ enzymes), B Complex (fuels stress-gut-liver-brain chain), Vitamin C (reduces cortisol, fortifies gut barrier), PS (blocks stress hormone release at pituitary), and Ashwagandha (normalizes stress response, protects liver). Managing stress isn't just mental health—it's disease prevention.

Did you know? — That swinging on swings and bouncing on trampolines activate the vagus nerve and trigger a cascade of feel-good brain chemicals, thereby improving health? The rhythmic motion releases serotonin (calms nervous system), dopamine builds with each predictable arc (pleasure and motivation), norepinephrine sharpens focus, and physical exertion releases endorphins.

At the peak — that brief weightlessness — all four can converge into frisson, those pleasurable chills. The health payoff: cortisol drops, parasympathetic nervous system activates (deep calm), mood lifts, anxiety reduces, sleep improves. Bonus: Improved lymphatic drainage, boost to brain health, and more resilient nervous system. It's a natural full-body neurochemical reset disguised as play — simultaneous relaxation and reward most activities can't produce.

One More Intriguing Cycle – The Longevity-Vitality Key

Looking back at all we've covered, disease prevention and recovery has been a major objective, but looking forward to achieving and maintaining vibrant health into the future is just as important if we want to optimize our life goals — in walks mitochondria.

The Process – CoQ10 and Mitochondrial Engine

Mitochondria are the furnace where everything we eat is converted into ATP, the currency of life. The next and last diagram traces the full cascade from dietary inputs to ATP production, mitochondrial quality control, and the self-renewing feedback loop that keeps the whole engine running. This process is so vital, it may well be the capstone of this entire course, so let's take a mini-dive one last time.

All macronutrients join at the Krebs cycle, where B3 and NAD^+ are vital. CoQ10, made from two streams — the mevalonate pathway (B vitamins, magnesium[1], liver enzymes) and tyrosine from diet or phenylalanine — then acts as the essential electron courier for ATP.

Taurine supports this process by enabling synthesis of the electron transport chain that make energy production possible for ATP. Yet, statins block the mevalonate pathway, muting CoQ10 synthesis as collateral damage to the very heart they're meant to protect.

---

[1]Magnesium's role goes far beyond CoQ10 synthesis. It's also required for glutathione — the cell's primary antioxidant — mitochondrial DNA repair, ATP stability, and membrane integrity. Deficiency at any of these points accelerates the very damage mitophagy exists to clean up.

CoQ10 also preserves nitric oxide availability, governing coronary dilation and oxygen delivery — which is why the heart, with the highest CoQ10 concentration of any tissue, feels its decline first.

CoQ10 is a core membrane antioxidant protecting cardiolipin for both energy production and mitophagy signaling, but astute research shows that its proper synthesis is also dependent on functioning cardiolipin itself via bi-directional feedback. Deplete it and you don't just slow ATP output; you silence the quality-control system that clears damaged mitochondria. (283. Tsui, 2019)

This is because cardiolipin, bolstered by CoQ10, plays the central role in signaling mitophagy cleanup as described in earlier sections, which in turn, generates new mitochondria triggered by the clearing out of damaged mitochondria. Incorporated into each new mitochondria, intact cardiolipin stabilizes the production of new CoQ10, thus completing the virtuous cycle. (284. Palikaras K, 2014)

Ideally, with the right nutrients, this could maximize natural biogenesis of CoQ10 and reduce the need for CoQ10 supplements as we age while preserving benefits from the cellular metabolic steps in the cycle that supplements may bypass like efficient mitophagy.

While PG-synthesized cardiolipin remains the nutritional linchpin for the whole process and omega-3s support fluidity in the membranes, research shows that the two critical nutrients necessary for optimizing mitophagy besides PG and exercise are urolithin A primarily from pomegranates and spermidine from a few key foods.

We've covered many nutrients and metabolic pathways, but perhaps the best and most elegant cycle was saved for last. Here, we see a uniquely illustrated process that promotes resilient longevity through robust mitophagy and mitochondrial regeneration.

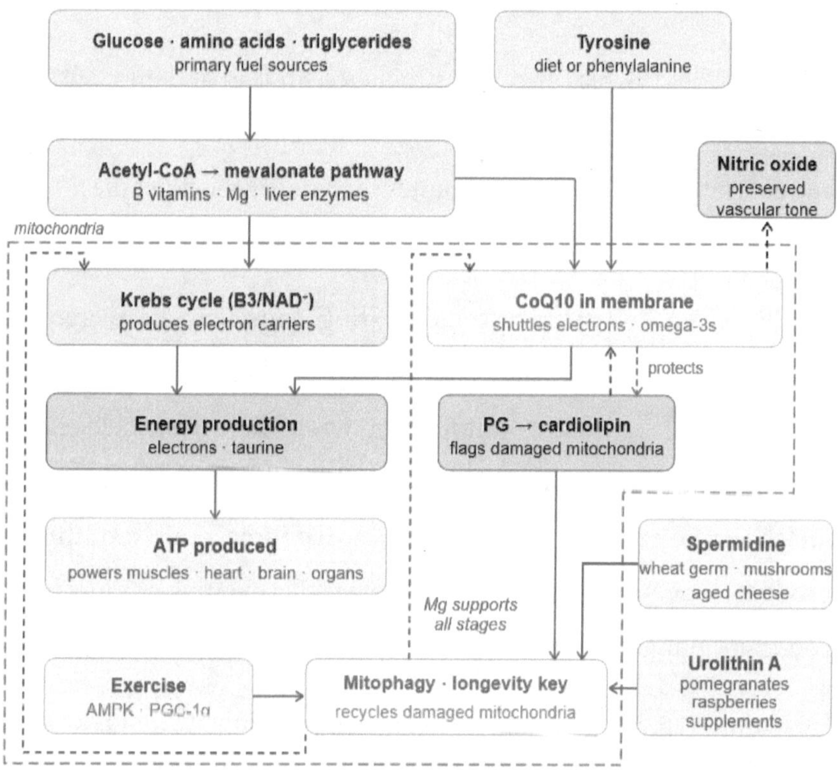

Figure 17. The Mitochondrial Engine © 2026

Mitophagy: The Longevity Key – Mitophagy, the body's engine of mitochondrial renewal, may be the single greatest factor in how we age, and explains why some older adults can outperform people half their age and be less dependent on certain supplements.

Mitophagy recycles damaged mitochondria and builds fresh replacements, driven by cardiolipin signals (protected by CoQ10), urolithin A from pomegranates, raspberries (supplement for 60% who can't convert); spermidine from wheat germ, mushrooms, aged cheese; and exercise — the only stimulus that activates cleanup and drives new growth simultaneously with nutrients.

CoQ10's AMPK bias also keeps metabolism lean and efficient, and because better mitochondria enable more exercise which drives better renewal; the gains compound into vitality most people assume belongs only to youth.

Quality sleep and time-restricted eating complete the picture — deep sleep is when the body executes much of its cellular housekeeping, including mitophagy, while time-restricted eating activates AMPK and suppresses mTOR, creating the same metabolic conditions that exercise, spermidine, and urolithin A produce. Together they ensure the renewal cycle runs not just when you train, but around the clock.

In summary, enhancing mitophagy is probably the single most effective tool for promoting longevity and vitality. Nutrients like tyrosine, B vitamins, Mg, omega-3s, and taurine are core strategic activators of cellular energy — while PG foods; spermidine from wheat germ, mushrooms, aged cheese; urolithin A from pomegranates, raspberries, and supplements; and vigorous exercise are the most potent drivers of mitophagy leading to resilient health.

The Top 12 Foods – A comprehensive Review

Previously, we covered the top seven foods for disease prevention and the top five nutrients for halting stress-induced disease at its source. Here, we'll zoom out to a broader range of healthy food groups that reflect a more comprehensive selection based not only on restoration, but also for maintaining a balanced metabolic spectrum of both AMPK and mTOR pathways for a resilient life going forward. AMPK still leads, but mTOR offers base support.

Fundamentally, apart from vegan diets, here's a best-take ranking based on all the data. The Top 12 core overlapping foods across the metabolic spectrum ranked by nutrient density, breadth of dietary impact, traditional wisdom, and realistic daily consumption:

CAVEAT: While Liver ranks #1 in nutrient density, consumption must be moderated and most people don't like the taste or texture.

1. Pastured Eggs – Perfect food; phospholipids, choline, vitamin A, lutein/zeaxanthin (eye health), vitamin D/E, B12, complete protein

2. Fermented dairy (kefir, yogurt, cheese) – Strongest longevity correlation; probiotics, C:15, CLA, butyrate, K2, complete protein, spermidine, calcium; balanced food for both AMPK and mTOR — grass-fed, A2, aged sheep/cow based milk for the best profile

3. Fish/Fatty fish (salmon, sardines, tuna, cod) – Omega-3 DHA with PLs (fatty fish); vitamin D, selenium, B12, B1 (esp. salmon), taurine (esp. yellowfin tuna), iodine, complete protein

4. Grass-fed red meat/poultry/shellfish (& bone broth) – Iron, taurine (esp. shellfish), tryptophan (for NAD, serotonin), zinc (esp. oysters), B vitamins (esp. B12), complete protein; glycine for bile, collagen for gut. Boiled/braised or marinate grilled meat (AGEs).

5. Leafy greens (especially spinach - cooked) – Source of #1 nutrient (magnesium); B9, vitamin K, vitamin C, calcium, non-heme iron

6. Nuts & seeds – Magnesium, zinc, B1; emphasize pistachios (melatonin, complete protein), Brazil nuts (selenium), walnuts (omega-3), pumpkin seeds (zinc), almonds

7. Cruciferous & fermented vegetables – Liver detoxification (sulforaphane); probiotics, vitamin C, K2; cabbage for gut healing

8. Olive oil – Biblical elixir of life; 75% monounsaturated fats, polyphenols, vitamins E & K; Mediterranean longevity

9. Whole grains (traditional prep) – Biblical staple; B vitamins, selenium, manganese, copper, magnesium, fiber, baseline energy

10. Fruits (especially pomegranates and berries) – Polyphenols, urolithin A, vitamin C; lower glycemic impact, longevity

11. Legumes/beans – Biblical staple and throughout the world; protein, fiber, resistant starch, folate, B1, iron, magnesium

12. Avocados – Monounsaturated and omega-3 fats, copper, potassium, magnesium, B vitamins, vitamins E & K, soluble fiber

Deeper Connections

Targeted approaches and bullet point actions are great tools to apply, but are only derived by synthesizing the big picture from a 30,000 foot view. If we step back from the details, certain profound patterns emerge linking everything we've explored in this book:

The Cascade Pattern: Metabolic disease flows like a stream from the top of a mountain to the ocean. When disorder begins here (often silently for years), it cascades eventually: fatty liver → unhealthy bile → leaky gut → bacterial toxins overwhelm liver → hepcidin-iron/HPA axis dysregulation → further gut/liver damage.

From this polluted source flow the rivers of most metabolic-related disease — downstream effects of the same upstream dysfunction. While targeted treatment may be needed, healing doesn't require fifty treatments for fifty symptoms. It requires purifying the source.

Divine Engineering: The human body reveals unmistakable intelligent design. The liver can regenerate from just 25% of its original tissue — why would God create such capacity unless it's so foundational to metabolic survival that losing it means total failure?

Elegant systems show brilliant engineering — built-in redundancy, molecules serving multiple functions, feedback loops self-correcting before disease cascades. Foods containing precise ratios impossible to replicate. From beginning to end, the amazing design unfolds. This isn't coincidence — it's providence!

Ancient Wisdom Vindicated: Modern science is rediscovering what the ancients knew through observation and divine revelation. TCM identified liver Qi stagnation as a root disease thousands of years before we understood fatty liver. Israelites ate fermented bread & dairy, diluted wine with meals, used olive oil as a dietary staple, and honored Sabbath rest — all practices modern research validates.

Dietary laws weren't arbitrary rules, but divine health wisdom. Cutting-edge research on the gut-brain axis and bile signaling is modern science discovering what the Ancients intuitively knew: the body is an integrated whole, food is medicine, mental-spiritual state affects health, and traditional foods surpass engineered alternatives.

Nutrition as Spiritual Practice: "Your bodies are temples of the Holy Spirit...honor God with your bodies" (1 Corinthians 6:19-20). Regardless of your beliefs, this isn't a metaphor, it's sacred care based on ancient wisdom. Habitually consuming ultra-processed foods and abusing our bodies through lifestyle and chronic stress aren't just poor health choices — it's desecrating the temple.

Conversely, nourishing ourselves with foods God has provided, honoring the body with activity and rest, addressing emotional wounds that manifest physically — that's honoring the temple. True healing addresses all these dimensions leading to cleaner health, sharper perspectives, and more resilience. Vibrant health isn't selfish vanity, but faithful stewardship of a divine gift, caring for the vehicle through which we accomplish our purpose.

Key Take-away Strategies

Understanding the big picture framing healthy perspectives along with specific remedies is invaluable, but bridging a middle ground for reviewing basic strategies to begin action is key as well. Thus, here are seven steps to consider from all the research we've covered, followed by core break-out findings to integrate and sum it all up:

Stop eating junk — This is a no-brainer. Cut out ultra-processed foods like vegetable oils, high fructose derivatives, and white flour in addition to all the other chemicals and additives in these foods.

Crush metabolic syndrome — Target proper nutrients/factors laid out in Part 5 and figures 7-11 for metabolic, brain, and CV health.

Disease prevention — Consider integrating the 7 primary AMPK foods listed in *Did you know* along with applicable supplements.

Evaluate stress-liver status —The gut-liver-brain axis is king, so heed the 5 core nutrients in *Did you know* for halting stress-disease

Exercise — Tailor a fitness regime based on the RE-EE exercise spectrum according to metabolic condition and future desired goals.

Healthy sustained diet — Consider the top 12 foods along with supplements as a foundation for a long-term strategic, healthy diet.

Enhance mitophagy for optimal health — Target key nutrients and exercise to support the mitochondrial engine for resilient longevity.

As we begin to wrap up, what are some core findings we can extract from this whole educational journey to help anchor our experience?

— The amazing design of the human body is vastly complex, but two main constructs seem to stand out — systems-level functioning and cellular-level performance — metabolic organ homeostasis and the mitochondrial engine anchored by mitophagy.

— The liver sits atop the metabolic organ pyramid as guardian of the gut-liver-brain triangle and master regulator of the whole pyramid. #5 nutrient PLs, #8 choline and #2 iron regulate and are regulated by a healthy liver for systemic homeostasis, but also play a role at the cellular level in cell membranes, nerves, and DNA.

— #1 Magnesium and #3 B vitamins are core nutrients essential across systemic and cellular processes — especially for energy and nervous system — aided by #4 sodium-potassium and #7 taurine, with iron and #6 zinc for nucleus/DNA to round out the top 8.

— The mitochondrial engine, supported by inputs above, but more specifically, CoQ10, omega-3s, PG/cardiolipin, urolithin A, spermidine, and exercise promote cellular resilience and longevity.

— Taking a holistic approach by enacting a smart diet, engaging in appropriate exercise, and reducing stress leading to better sleep along with traditional modes like acupuncture, heat/cold exposure, sun/redlight therapy, and grounding; all the while respecting our temple and the one who gave it, can lead us toward resilient health.

In Closing

When my health crisis struck a few years ago, it took me on a path I didn't expect — one that would transform my health and ultimately inspire me to write this book. No doubt, new discoveries will continue to surface as will additional dots to connect, but the healing path prescribed in this book worked because it addressed root causes. My story isn't unique — this cascade affects millions.

Fast forward to now and I feel like I've turned back the clock several years, regaining my health, mental clarity, metabolic resilience, and youthful vigor; but it all started with that still small voice saying:
"Here son, take this atrocious-looking lemon and make some awesome lemonade, not just for yourself, but to help others as well".

What began as a desperate search for personal healing became a comprehensive map I hope can guide others. Whether you're facing crisis or optimizing vitality, the principles are the same. Start where you are, take the next step by honoring the temple you've been given, and trust not only the process, but the one who gave it.

I know we all face challenges, but I do humbly pray this booked has blessed and encouraged you by offering insights and inspiration as a launch pad for reclaiming your own health. Remember, wisdom is more precious than rubies. May you find that wisdom, apply it faithfully, and experience the vibrant health that is your birthright. Discovering abundant health is not always an easy path, but it's profoundly worth walking and I'm grateful you walked it with me.

Afterword

Throughout this book, the liver has rightfully claimed center stage — master regulator of the metabolic organ pyramid, orchestrator of the gut-liver-brain triangle, commander of bile synthesis and systemic homeostasis. When it falters, disease cascades. When it heals, conditions for vitality return.

But the nervous system has been present throughout as well — woven into every section, though perhaps not always named as such. Magnesium and B vitamins, our #1 and #3 core nutrients, are architects of nervous system function.

The gut-liver-brain axis is fundamentally neurological, mediated by the vagus nerve connecting organs, regulating inflammation, and coordinating healing. Stress management, sleep, exercise, even swinging and bouncing — all activate parasympathetic pathways and vagal tone, shifting the body from survival into restoration.

We've covered this foundation. Now, emerging science is revealing just how profoundly the nervous system directs everything we've discussed — and extending our understanding in remarkable ways.

What's Emerging

In 2025, researchers showed that vagus nerve stimulation reduces rheumatoid arthritis not by masking pain, but by modulating the immune system's inflammatory reflex at its source — disease activity and joint damage both declined. (285. Tesser JRP, 2025)

Scientists discovered that maintaining vagal connections to the heart protects cardiac muscle and slows aging regardless of heart rate, confirming the nerve's signaling primacy over mechanical factors. (286. Anar Dushpanova, 2025)

The mitochondrial engine we explored in Part 7 doesn't operate in isolation. Mitochondrial-nervous system crosstalk is proving vital for cellular health. When this communication falters, neurons lose plasticity, oxidative stress surges, and neuroinflammation cascades — especially devastating in the brain. Chronic stress doesn't just elevate cortisol; it damages mitochondria in neurons, creating a vicious cycle of energy depletion and nervous system dysregulation.
(287. Kelly C, 2024) (288. Fagundes CP, 2025)

Perhaps most remarkably, microglial cells form tunneling nanotubes to transfer healthy mitochondria directly into struggling neurons, actively rescuing them from oxidative damage. The nervous system doesn't just coordinate function — it intervenes to repair damage. (289. Scheiblich H, 2024)

The Network Extends

What we called the gut-liver-brain triangle could be more accurately called the gut-liver-NS network The vagus nerve coursing through the torso. The enteric nervous system — that "second brain" with more neurons than the spinal cord. The autonomic nervous system balancing fight-or-flight with rest-and-repair. Peripheral nerves and glial cells coordinating across tissues.

The liver may be the metabolic master, but the nervous system is the conductor — directing when to burn fat or store it, when to activate mitophagy or conserve energy, when to fight infection or heal tissue. If mitochondria are the engine and the liver the fuel refinery, the nervous system as a whole is the wiring, throttle, and diagnostic system coordinating it all.

When these systems function optimally, a virtuous cycle emerges: Stress-managed, choline-supported liver mitochondria → systemic nutrient synthesis → enhanced mitochondrial engine body-wide → resilient nervous system → reduced stress and toxin load → healthier organs, lower inflammation → reduced liver burden → efficient regeneration, sustained by intentional health practices → virtuous cycle of abundant health.

This isn't revision. It's validation. Everything we've covered — organ interconnections, nutrient synergies, ancient wisdom, metabolic restoration — stands firm. Emerging research is simply illuminating deeper layers of the same elegant design.

Where This Leads

The foundation you've learned equips you to understand what's unfolding. Emerging research shows that mitochondrial aging in neurons drives cognitive decline, yet nervous system modulation may reverse it through enhanced mitophagy. The nervous system regulates not just movement, but immune function, metabolic rate, hormone release, gene expression — the whole symphony.

Your entry point might have been liver dysfunction like mine, or chronic fatigue, anxiety, brain fog, or metabolic syndrome. Wherever you started, the threads lead to the same truth: abundant health flows from seamless communication between metabolic, immune, hormonal, and neurological systems — all coordinated by that vast, adaptive nervous system network.

The research is accelerating. New discoveries validate what ancient wisdom intuited and what we've explored together. The dots connect further with each study, revealing a design more intricate and beautiful than we imagined.

So consider this not an ending, but an invitation. You've learned how organs interconnect, how nutrients fuel function, how the liver orchestrates metabolism, how mitophagy sustains longevity. You've built the foundation for understanding what comes next.

And perhaps, as the science unfolds and the picture clarifies, we'll explore that expanding frontier together.

May you discover not just abundant health, but profound wonder at the elegance of the design — and peace in knowing your body, in all its complexity, works exactly as it was meant to.

References

1. Axe, D. (2017). *Stop Eating These Oils Immediately.* Retrieved from Dr. Josh Axe: https://www.youtube.com/watch?v=J6BigpBa_9g

2. Berg, D. (2022). *The Food That is More DEADLY Than Sugar.* Retrieved from Dr. Eric Berg DC: https://www.youtube.com/watch?v=V2_Jmd6mCUQ

3. Eckberg, D. (2021). *Ten Cooking Oils...The Good, Bad and Toxic.* Retrieved from YouTube: https://www.youtube.com/watch?v=pljQrjiDC9Q

4. Berg, D. (2019). *A Surprising Way To Cleanse Fatty Liver.* Retrieved from Dr. Eric Berg DC: https://www.youtube.com/watch?v=weZ_BoMptLI

5. Ekberg, D. (2019). *How To Reverse Fatty Liver Disease.* Retrieved from Dr. Sten Ekberg: https://www.youtube.com/watch?v=AgSjlQIlwIo

6. Pam, D. (2018). *Whole Grains Reduce Inflammation.* Retrieved from Pamela Popper: https://www.youtube.com/watch?v=t_dtpzfMmXE

7. Berg, D. (2019). *What's Worse: White Sugar vs White Flour?* Retrieved from Dr. Eric Berg DC: https://www.youtube.com/watch?v=g_pax5an8B4

8. Gundry, D. (2022). *Is MEAT Healthy For You.* Retrieved from The Dr. Gundry Podcast: https://www.youtube.com/watch?v=XXTH7fQzK4Y

9. Berg, D. (2020). *What Is Bile Reflux.* Retrieved from Dr. Eric Berg DC: https://www.youtube.com/watch?v=K63PQzIfmFE

10. Axe, D. (2016). *How To Overcome Gastritis.* Retrieved from Dr. Josh Axe: https://www.youtube.com/watch?v=pLwTXMDCdV0

11. Gundry, D. (2021). *Ultimate polyphenol guide for better gut health.* Retrieved from Gundry MD: https://www.youtube.com/watch?v=wF8Y3fTXsG8

12. Berg, D. (2020). *What Is Gastritis?* Retrieved from Dr. Eric Berg DC: https://www.youtube.com/watch?v=B9SlvZ2wqUI

13. Berg, D. (2019). *A Surprising Way To Cleanse Fatty Liver.* Retrieved from Dr. Eric Berg DC: https://www.youtube.com/watch?v=weZ_BoMptLI

14. Ekberg, D. (2022). *#1 Absolute Best FATTY LIVER DETOX.* Retrieved from Dr. Sten Ekberg: https://www.youtube.com/watch?v=u4i48FBwfs4

15. Berg, D. (2018). *10 Things That Increase Bile Salts.* Retrieved from Dr. Eric Berg DC: https://www.youtube.com/watch?v=h64_7DYf5_0

16. Brian St. Pierre, M. R. (n.d.). *Settling the great grain debate.* Retrieved from Precision Nutrition: https://www.precisionnutrition.com/grain-wheat-debate

17. Axe, D. (2017). *Top 15 Anti-Inflammatory Foods.* Retrieved from Dr. Josh Axe: https://www.youtube.com/watch?v=SVIDLu7s4HE

18. Gundry, D. (2017). *Think you can't eat dairy? Think again.* Retrieved from Gundry MD: https://www.youtube.com/watch?v=bQ28aI4e1LU

19. Berg, D. (2018). *The Amazing Magnesium Benefits.* Retrieved from Dr. Eric Berg DC: https://www.youtube.com/watch?v=rq2pcEQhk_w

20. QNL. (n.d.). *Plant Multi-Vitamin, Quantum.* Retrieved from Quantum Nutrition Labs: https://qnlabs.com/plant-multi-vitamin-quantum.html

21. Livingood, D. (2022). *7 Impressive Benefits of Vitamin C.* Retrieved from Dr. Livinggood: https://www.youtube.com/watch?v=XhzwdEFA1Ek

22. Axe, D. (2023). *Top Zinc Benefits to Improve Health & How to Regulate Levels.* Retrieved from Dr. Axe: https://draxe.com/nutrition/zinc-benefits/

23. Osborne, D. (2019). *L-Glutamine - Who Benefits From Supplementation?* Retrieved from Peter Osborne: https://www.youtube.com/watch?v=Ai7NtL4HBLc&t=235s

24. DeLauer, T. (2024). *5 Reasons Taurine is the #1 Longevity Supplement You Can Legally Take.* Retrieved from Thomas DeLauer: https://www.youtube.com/watch?v=IhvyqGUSqo4

25. Childs, D. (2023). *Take Tyrosine For Better Adrenal, Thyroid & Brain Health.* Retrieved from Dr. Westin Childs: https://www.youtube.com/watch?v=0Pt9ZpDHUw4

26. Berg, D. (2024). *The #1 Collagen Tip for Amazing Hair, Nails, & Skin.* Retrieved from Dr. Eric Berg DC: https://www.youtube.com/watch?v=DEiybCei-ls&t=26s

27. Berg, D. (2021). *The Benefits of TUDCA.* Retrieved from Dr. Eric Berg DC: https://www.youtube.com/watch?v=boTzigj646Q

28. Berg, D. (2023). *The Mind-Blowing Benefits of Berberine.* Retrieved from Dr. Eric Berg DC: https://www.youtube.com/watch?v=QX6gCRJZ0dk

29. Spa-Doctor, T. (2019). *The Liver's Connection to Emotions and Health with Dr. Ameet Aggarwal.* Retrieved from The Spa Doctor Podcast: https://www.youtube.com/watch?v=EyEk_CQwoGg

30. Shaw, B. (2019). *When Trauma Gets Stuck in the Body.* Retrieved from Pschology Today: https://www.psychologytoday.com/us/blog/in-the-body/201910/when-trauma-gets-stuck-in-the-body

31. Jin-Yong Joung, J.-H. C.-H.-H.-G. (2019). A literature review for the mechanisms of stress-induced liver injury. *Brain and Behavior, 9*(3). doi:https://doi.org/10.1002/brb3.1235

32. Scott JH, M. M. (2023). *Physiology, Aldosterone.* StatPearls Publishing. Retrieved from StatPearls: https://www.ncbi.nlm.nih.gov/books/NBK470339/

33. Cleveland Clinic. (2021). *Parathyroid Hormone.* Retrieved from Cleveland Clinic: https://my.clevelandclinic.org/health/articles/22355-parathyroid-hormone

34. DiNicolantonio, D. J. (2021). *Why You Need More Salt in Your Diet.* Retrieved from Dr. Eric Berg DC: https://www.youtube.com/watch?v=RV3o0VWfHuQ

35. Medical News Today. (2016). *Low salt intake may raise risk of heart attack, stroke, and death.* Retrieved from Medical News Today: https://www.medicalnewstoday.com/articles/310447

36. Berg, D. (2019). *The Sodium-Potassium Ratio is More Important Than a Low Sodium Diet.* Retrieved from Dr. Eric Berg DC: https://www.youtube.com/watch?v=z4MFnrMpzBA

37. Li S, X. X. (2023). Hydration Status in Older Adults: Current Knowledge and Future Challenges. *NJtrients, 15*(11). doi:https://doi.org/10.3390/nu15112609

38. Wang J, L. F. (2023). The organum vasculosum of the lamina terminalis and subfornical organ: regulation of thirst. *Frontiers in Neuroscience, 17.* doi:https://doi.org/10.3389/fnins.2023.1223836

39. Takashi Matsuda, T. Y. (2020). Distinct CCK-positive SFO neurons are involved in persistent or transient suppression of water intake. *Nature*(5692). doi:https://doi.org/10.1038/s41467-020-19191-0

40. Sangkhae V, N. E. (2017). Regulation of the Iron Homeostatic Hormone Hepcidin. *Advances in Nutrition, 8*(1). doi:https://doi.org/10.3945/an.116.013961

41. Subramaniam, G. R. (2017). The liver in regulation of iron homeostasis. *American Journal of Physiology-Gastrointestinal and Liver Physiology, 313*(3). doi:https://journals.physiology.org/doi/full/10.1152/ajpgi.00004.2017

42. Szklarz M, G.-N. K.-S. (2022). Can Iron Play a Crucial Role in Maintaining Cardiovascular Health in the 21st Century? *International Journal of Environmental Research and Public Health, 19*(19). doi: https://doi.org/10.3390/ijerph191911990

43. Angove, D. M. (2023). *Blood Vitality*. Retrieved from Optimize Iron: https://optimizeiron.com/the-connection-between-iron-deficiency-and-copper/

44. Xie, L. e. (2022). Downregulation of hepatic ceruloplasmin ameliorates NAFLD via SCO1-AMPK-LKB1 complex. *Cell Reports, 41*(3). Retrieved from https://www.cell.com/cell-reports/fulltext/S2211-1247(22)01348-1?_returnURL=https%3A%2F%2Flinkinghub.elsevier.com%2Fretrieve%2Fpii%2FS2211124722013481%3Fshowall%3Dtrue

45. Jillian Levy, C. (2018). *AMPK: The Energy-Boosting, Hormone-Balancing Benefits of AMPK*. Retrieved from Dr. Axe: https://draxe.com/nutrition/ampk/

46. Diana Vara-Ciruelos, F. M. (2019). The strange case of AMPK and cancer: Dr Jekyll or Mr Hyde? *Open Biology, 9*(7). doi:https://doi.org/10.1098/rsob.190099

47. Yucan Dai, H. T. (2021). The Crucial Roles of Phospholipids in Aging and Lifespan Regulation. *Frontiers in Physiology, 12*. doi:https://doi.org/10.3389/fphys.2021.775648

48. Heden, T. D. (2016). Looking Beyond Structure: Membrane Phospholipids of Skeletal Muscle Mitochondria. *Trends in Endocrinology & Metabolism, 27*(8). doi:http://dx.doi.org/10.1016/j.tem.2016.05.007

49. Osman C, V. D. (2011). Making heads or tails of phospholipids in mitochondria. *Journal of Cell Biology*. doi:https://doi.org/10.1083/jcb.201006159

50. Choi J, Y. T. (2017). Comprehensive analysis of phospholipids in the brain, heart, kidney, and liver: brain phospholipids are least enriched with polyunsaturated fatty acids. *Molecular and Cellular Biochemistry, 442*. doi:https://doi.org/10.1007/s11010-017-3203-x

51. Wnorowski, D. T. (2021). *Phosphatidylcholine vs Phosphatidylserine: Which Phospholipid Is Right for You?* Retrieved from BodyBio: https://bodybio.com/blogs/blog/phosphatidylserine-the-death-of-a-cell

52. Chen, W. C. (2018). Phosphatidylglycerol Incorporates into Cardiolipin to Improve Mitochondrial Activity and Inhibits Inflammation. *Scientific Reports, 8*(4919). doi:https://doi.org/10.1038/s41598-018-23190-z

53. Christie, W. W. (2024). *Phosphatidic Acid, Lysophosphatidic Acid and Related Lipids.* Retrieved from The Lipid Web: https://www.lipidmaps.org/resources/lipidweb/lipidweb_html/lipids/complex/pa/index.htm

54. ANS, 2. (2024). *Autonomic nervous system.* Retrieved from Wikipedia: https://en.wikipedia.org/wiki/Autonomic_nervous_system

55. Hopkins, J. (n.d.). *Brain Anatomy and How the Brain Works.* Retrieved from John Hopkins Medicine: https://www.hopkinsmedicine.org/health/conditions-and-diseases/anatomy-of-the-brain

56. Challenged, N. (n.d.). *Know Your Brain: Amygdala.* Retrieved from Neuroscientifically Challenged: https://neuroscientificallychallenged.com/posts/know-your-brain-amygdala

57. Study Smarter. (n.d.). *Fight-or-flight-response.* Retrieved from Study Smarter: https://www.studysmarter.co.uk/explanations/psychology/social-context-of-behaviour/fight-or-flight-response/

58. Lifestyle Desk. (n.d.). *Know Your Body.* Retrieved from Indian Express: https://indianexpress.com/article/lifestyle/health/know-your-body-human-brain-made-60-per-cent-fat-8648797/

59. Sherzai, D. (n.d.). *Healthy fats for the brain: myths, science, and diets.* Retrieved from The Brain Docs: https://thebraindocs.com/brain-healthy-fats/

60. Elizabeth Calzada, O. O. (2016). Chapter Two - Phosphatidylethanolamine Metabolism in Health and Disease. *International Review of Cell and Molecular Biology, 321.* doi:https://doi.org/10.1016/bs.ircmb.2015.10.001

61. Palafox-Sánchez, V. (2021). The interaction between brain and liver regulates lipid metabolism in the TBI pathology. *Biochimica et Biophysica Acta (BBA) - Molecular Basis of Disease, 1867*(4). doi:https://doi.org/10.1016/j.bbadis.2021.166078

62. Marano, H. E. (2018). *The Cognitive Benefits of Vitamin C.* Retrieved from Psychology Today: https://www.psychologytoday.com/us/articles/201801/the-cognitive-benefits-vitamin-c

63. Adães, S. (2018). *What is Glutamate.* Retrieved from Qualia: https://www.qualialife.com/what-is-glutamate

64. Keenan A. Walker, B. N. (2019). Understanding the Role of Systemic Inflammation in Alzheimer's Disease. *ACS Chemical Neuroscience, 10*(8). Retrieved from https://pubs.acs.org/doi/10.1021/acschemneuro.9b00333

65. Grimble PHD, R. F. (1998). Nutritional modulation of cytokine biology. *Nutrition, 14*(7-8). Retrieved from https://doi.org/10.1016/S0899-9007(98)00010-0

66. Cortese K, G. M. (2023). Interactions between Glycine and Glutamate through Activation of Their Transporters in Hippocampal Nerve Terminals. *Biomedicines, 11*(12). Retrieved from https://doi.org/10.3390/biomedicines11123152

67. Schmoe, J. (n.d.). *Nutrients to help repair your blood-brain barrier.* Retrieved from The Functional Neurology Center: https://thefnc.com/research/nutrients-to-help-repair-your-blood-brain-barrier/

68. Kubala, J. (2024). *6 Best Evidence-Based Supplements for Brain Fog.* Retrieved from Healthline: https://www.healthline.com/nutrition/vitamins-for-brain-fog

69. National University. (n.d.). *Best Brain Foods for Studying.* Retrieved from National University: https://www.nu.edu/blog/best-brain-foods-for-studying/

70. Ignite Healthwise. (2024). *Electrical System of the Heart.* Retrieved from MyHealth.Alberta.ca: https://myhealth.alberta.ca/Health/Pages/conditions.aspx?hwid=te7147abc

71. Nakagawa, N. K. (2019). Atrial and brain natriuretic peptides: Hormones secreted from the heart. *111.* doi:10.1016/j.peptides.2018.05.012

72. Cleveland Clinic. (2022). *Renin-Angiotensin-Aldosterone System (RAAS).* Retrieved from Cleveland Clinic: https://my.clevelandclinic.org/health/articles/24175-renin-angiotensin-aldosterone-system-raas

73. Palit, K. (2014). Vascular Calcification in Chronic Kidney Disease: Role of Disordered Mineral Metabolism. *Bentham Science, 20*(37). doi:10.2174/138161282066140212194926

74. Yahagi, K. L. (2016). Pathology of Human Coronary and Carotid Artery Atherosclerosis and Vascular Calcification in Diabetes Mellitus. *Arteriosclerosis, Thrombosis, and Vascular Biology, 37*(2). doi:https://doi.org/10.1161/ATVBAHA.116.306256

75. Ivanova, M. M. (2017). Small Dense Low-Density Lipoprotein as Biomarker for Atherosclerotic Diseases. *Oxidative Medicine and Cellular Longevity.* doi:https://doi.org/10.1155/2017/1273042

76. Froyen, E. (2021). The effects of fat consumption on low-density lipoprotein particle size in healthy individuals: a narrative review. *Lipids in Health and Disease.* doi:https://doi.org/10.1186/s12944-021-01501-0

77. Berg, D. (2024). *Bad Cholesterol LDL Is NOT Bad.* Retrieved from Dr. Berg: https://www.drberg.com/blog/bad-cholesterol-ldl-is-not-bad

78. Choi, L. P. (2018). Zinc deficiency and cellular oxidative stress: prognostic implications in cardiovascular diseases. *Acta Pharmacologica Sinica.* Retrieved from https://www.nature.com/articles/aps201825

79. Sternberg, e. a. (2023). Ketogenic diets composed of long-chain and medium-chain fatty acids induce cardiac fibrosis in mice. *Molecular Metabolism, 72.* doi:https://doi.org/10.1016/j.molmet.2023.101711

80. Surdu AM, P. O. (2021). Vitamin D and Its Role in the Lipid Metabolism and the Development of Atherosclerosis. *Biomedicines, 9*(2). doi:https://doi.org/10.3390/biomedicines9020172

81. Sunlight Institute. (2017). *Sun Exposure and Vitamin D Reduce Cholesterol levels.* Retrieved from Sunlight Institute: https://sunlightinstitute.org/sun-exposure-vitamin-d-cholesterol/

82. Cytoplan. (2017). *Research surrounding the benefits of CoQ10.* Retrieved from Cytoplan Blog: https://blog.cytoplan.co.uk/research-surrounding-benefits-coq10/

83. Jong CJ, S. P. (2021). The Role of Taurine in Mitochondria Health: More Than Just an Antioxidant. *Molecules, 26*(16). doi:https://doi.org/10.3390/molecules26164913

84. Amino Science. (2018). *Amino Acids for the Heart: How Amino Acids Help Lower Blood Pressure.* Retrieved from The Amino Co: https://aminoco.com/blogs/amino-acids/amino-acids-for-the-heart

85. Qardio. (2017). *HEALTHY HEART BLOG.* Retrieved from Qardio: https://www.qardio.com/healthy-heart-blog/ten-nutrients-heart-health-youve-probably-never-heard/

86. AJmera, R. (2025). *17 Incredibly Heart-Healthy Foods.* Retrieved from Healthline: https://www.healthline.com/nutrition/heart-healthy-foods

87. Cleveland Clinic. (2024). *Respiratory System.* Retrieved from Cleveland Clinic: https://my.clevelandclinic.org/health/body/21205-respiratory-system

88. Kuba K, I. Y. (2006). Angiotensin-converting enzyme 2 in lung diseases. *Current Opinion in Pharmacology, 6*(3). doi:https://doi.org/10.1016/j.coph.2006.03.001

89. Taskin E, G. C. (2017). *Local Renin-Angiotensin System at Liver and Crosstalk with Hepatic Diseases.* IntechOpen. Author. doi:DOI: 10.5772/65919

90. Agassandian M, M. R. (2013). Surfactant phospholipid metabolism. *Biochimica et Biophysica Acta (BBA) - Molecular and Cell Biology of Lipids, 1831*(3). doi:https://doi.org/10.1016/j.bbalip.2012.09.010

91. Alves E, D. M. (2018). Polar Lipids from Olives and Olive Oil: A Review on Their Identification, Significance and Potential Biotechnological Applications. *Foods.* doi:https://doi.org/10.3390/foods7070109

92. Breitzig MT, A. M. (2018). Thyroid hormone: a resurgent treatment for an emergent concern. *American Journal of Physiology-Lung Cellular and Molecular Physiology, 315*(6). doi:https://doi.org/10.1152/ajplung.00336.2018

93. Anathy V, L. K. (2018). Reducing protein oxidation reverses lung fibrosis. *Nature Medicine.* doi:https://doi.org/10.1038/s41591-018-0090-y

94. Tuzova M, J. J.-B. (2014). Inhibiting lung lining fluid glutathione metabolism with GGsTop as a novel treatment for asthma. *Frontiers in Pharmacology, 5.* doi:https://doi.org/10.3389/fphar.2014.00179

95. Landry, J. (2025). *13 Best Vitamins and Supplements for Lung Health.* Retrieved from Respiratory Therapy Zone: https://www.respiratorytherapyzone.com/best-vitamins-for-lungs/

96. J Clin Invest. (1980). Renal metabolism of amino acids and ammonia in subjects with normal renal function and in patients with chronic renal insufficiency. doi: https://doi.org/10.1172/JCI109771

97. Friedman, T. M. (2003). *The Importance of the Adrenal Cortex Hormones Cortisol and Aldosterone.* Retrieved from Cushing's Support & Research Foundation: https://csrf.net/doctors-articles/steroid-replacement/the-importance-of-the-adrenal-cortex-hormones-cortisol-and-aldosterone/

98. Umbro, B. F. (2021). Nonalcoholic Fatty Liver Disease and the Kidney: A Review. *Biomedicines, 9*(10). doi:https://doi.org/10.3390/biomedicines9101370

99. Florens N, C. C. (2016). Modified Lipids and Lipoproteins in Chronic Kidney Disease: A New Class of Uremic Toxins. *Toxins, 8*(12). doi:https://doi.org/10.3390/toxins8120376

100. Yoon YM, K. S. (2019). TUDCA-treated chronic kidney disease-derived hMSCs improve therapeutic efficacy in ischemic disease via PrPC. *Redox Biology, 22.* doi:https://doi.org/10.1016/j.redox.2019.101144

101. Fernandez, S. (2019). *Reversing polycystic kidney disease.* University of California, News. Retrieved from https://www.universityofcalifornia.edu/news/reversing-polycystic-kidney-disease

102. Den Hartogh D, T. E. (2019). Health Benefits of Resveratrol in Kidney Disease: Evidence from In Vitro and In Vivo Studies. *Nutrients, 11*(7). doi:https://doi.org/10.3390/nu11071624

103. Picard K, B. S. (2020). Dietary Potassium Intake and Risk of Chronic Kidney Disease Progression in Predialysis Patients with Chronic Kidney Disease: A Systematic Review. *Advances in Nutrition, 11*(4). doi:https://doi.org/10.1093/advances/nmaa027

104. Chu SH, Y. D. (2021). Effect of resveratrol on the repair of kidney and brain injuries and its regulation on klotho gene in d-galactose-induced aging

mice. *Bioorganic & Medicinal Chemistry Letters, 40.* doi:https://doi.org/10.1016/j.bmcl.2021.127913

105. Buchanan S, C. E. (2020). Klotho, Aging, and the Failing Kidney. *Frontiers in Endocrinology, 11.* doi:https://doi.org/10.3389/fendo.2020.00560

106. Wu S, C. Y. (2022). Adherence to Mediterranean Diet and Soluble Klotho Level: The Value of Food Synergy in Aging. *Nutrients, 14*(19). doi:https://doi.org/10.3390/nu14193910

107. Cleveland Clinic. (2024). *Pancreas.* Retrieved from Cleveland Clinic: https://my.clevelandclinic.org/health/body/21743-pancreas

108. PanCan. (n.d.). *Pancreatic Enzymes.* Retrieved from Pancreatic Cancer Action Network: https://pancan.org/facing-pancreatic-cancer/living-with-pancreatic-cancer/diet-and-nutrition/pancreatic-enzymes/

109. You and Your Hormones. (2021). *Somatostatin.* Retrieved from You and Your Hormones: https://www.yourhormones.info/hormones/somatostatin/

110. Babic T, T. A. (2016). *Neural Control of the Pancreas.* doi:DOI: 10.3998/panc.2016.27

111. Abramowitz, J. M. (2016). *Living without a pancreas: Is it possible?* Retrieved from UT Southwestern Medical Center: https://utswmed.org/medblog/living-without-a-pancreas/

112. Rutter, G. (2017). GABA signaling: A route to new pancreatic β cells. *Cell Research, 27.* doi:https://doi.org/10.1038/cr.2017.20

113. Vlavcheski F, O. E. (2022). Effects of Berberine against Pancreatitis and Pancreatic Cancer. *Molecules, 2723.* doi:https://doi.org/10.3390/molecules27238630

114. Zhang CL, W. J. (2021). Nonalcoholic fatty pancreas disease: An emerging clinical challenge. *World Journal of Clinical Cases, 9*(23). doi:doi: 10.12998/wjcc.v9.i23.6624

115. Ulasoglu C, T. Z. (2021). Does Nonalcoholic Pancreatic Steatosis Always Correlate with Nonalcoholic Fatty Liver Disease? *Clinical and Experimental Gastroenterology*. doi:https://doi.org/10.2147/CEG.S317340

116. Spero, D. B. (2013). *Healing Leaky Livers*. Retrieved from Diabetes Self-Management: https://www.diabetesselfmanagement.com/blog/healing-leaky-livers/

117. Liver Doctor. (n.d.). *Liver Doctor*. Retrieved from Do You Have A Fatty Pancreas?: https://www.liverdoctor.com/do-you-have-a-fatty-pancreas/

118. Juechter, S. M. (2022). *10 Amazing Foods to Heal Your Pancreas*. Retrieved from eMediHealth: https://www.emedihealth.com/nutrition/healing-foods-for-pancreas

119. Cleveland Clinic. (2021). *Spleen*. Retrieved from Cleveland Clinic: https://my.clevelandclinic.org/health/body/21567-spleen

120. Tarantino G, S. A. (2013). Liver-spleen axis: Intersection between immunity, infections and metabolism. *World Journal of Gastroenterology, 19*(23). doi:doi: 10.3748/wjg.v19.i23.3534

121. Wu, X. (1998). Current concept of Spleen-Stomach theory and Spleen deficiency syndrome in TCM. *World Journal of Gastroenterology, 4*(1). doi:doi: 10.3748/wjg.v4.i1.2

122. Cleveland Clinic. (2023). *Lymphatic System*. Retrieved from Cleveland Clinic: https://my.clevelandclinic.org/health/body/21199-lymphatic-system

123. Gardner, A. (2022). *How to Stimulate Thymus Function Naturally*. Retrieved from Livestrong: https://www.livestrong.com/article/548758-how-to-increase-the-thymus-function/

124. Keown, S. (2018). *How the immune system's key organ regenerates itself*. Retrieved from Fred Hutchinson Cancer Center: https://www.fredhutch.org/en/news/center-news/2018/01/thymus-regeneration.html

125. Deak, T. (2008). Immune Cells and Cytokine Circuits: Toward a Working Model for Understanding Direct Immune-to-Adrenal Communication Pathways. *Endocrinology, 149*(4). doi:https://doi.org/10.1210/en.2008-0170

126. Stanford, University Medical Center. (2012). *How stress can boost immune system.* Retrieved from ScienceDaily: https://www.sciencedaily.com/releases/2012/06/120621223525.htm

127. Gardner, A. (2022). *How to Stimulate Thymus Function Naturally.* Retrieved from Livestrong: https://www.livestrong.com/article/548758-how-to-increase-the-thymus-function/

128. Krishnamurthy HK, R. S. (2021). Effect of Micronutrients on Thyroid Parameters. *Thyroid Research.* doi: https://doi.org/10.1155/2021/1865483

129. Childs, W. D. (2025). *L-Tyrosine and How it Benefits Thyroid & Adrenal Function: Should you Use it?* Retrieved from Dr. Childs: https://www.restartmed.com/l-tyrosine/

130. Hedberg, D. N. (n.d.). *Ferritin and Hypothyroidism.* Retrieved from https://drhedberg.com/ferritin-hypothyroidism/

131. Wikipedia. (n.d.). *Adrenal Gland.* Retrieved from Wikipedia: https://en.wikipedia.org/wiki/Adrenal_gland

132. Axe, D. J. (2023). *What Is Adrenal Fatigue? Steps to Overcome It Naturally.* Retrieved from Dr. Axe: https://draxe.com/health/adrenal-fatigue/

133. Sensoy, I. (2021). A review on the food digestion in the digestive tract and the used in vitro models. *Current Research in Food Science, 4.* doi:https://doi.org/10.1016/j.crfs.2021.04.004

134. Kiela PR, G. F. (2016). Physiology of Intestinal Absorption and Secretion. *Best Practice & Research Clinical Gastroenterology, 30*(2). doi:https://doi.org/10.1016/j.bpg.2016.02.007

135. Sprouse, S. (n.d.). *Your Guide to How Nutrients are Absorbed by the Body.* Retrieved from Ask The Scientists: https://askthescientists.com/nutrient-absorption/

136. GrassrootsHealth. (2021). *The Relationship between Vitamin D, Sunlight, and the Digestive System.* Retrieved from GrassrootsHealth Nutrient Research Institute: https://www.grassrootshealth.net/blog/relationship-vitamin-d-sunlight-digestive-system/#

137. Scaccia, A. (2024). *Everything You Need to Know About Serotonin.* Retrieved from Healthline: https://www.healthline.com/health/mental-health/serotonin

138. Larabi AB, M. H. (2023). Bile acids as modulators of gut microbiota composition and function. *Gut Microbes, 15*(1). doi:https://doi.org/10.1080/19490976.2023.2172671

139. Wang HH, P. P. (2019). Update on the Molecular Mechanisms Underlying the Effect of Cholecystokinin and Cholecystokinin-1 Receptor on the Formation of Cholesterol Gallstones. *Current Medicinal Chemistry, 26*(19). Retrieved from Cleveland Clinic.

140. Yeo XY, T. L. (2023). Liver's influence on the brain through the action of bile acids. *Frontiers in Neuroscience, 17.* doi:https://doi.org/10.3389/fnins.2023.1123967

141. Harvard Health. (2019). *Stress and the sensitive gut.* Harvard Medical School, Harvard Health Online. Harvard Health Publishing. Retrieved from Harvard Health Online: https://www.health.harvard.edu/newsletter_article/stress-and-the-sensitive-gut

142. Axe, D. (2024). *Bone Broth Benefits for Joints, Skin and Gut Health.* Retrieved from Dr. Axe: https://draxe.com/nutrition/bone-broth-benefits/

143. Rao JN, W. J. (2010). Luminal Nutrients and Microbes in Gut Mucosal Growth. In W. J. Rao JN, *Regulation of Gastrointestinal Mucosal*

Growth. Morgan & Claypool Life Sciences. Retrieved from https://www.ncbi.nlm.nih.gov/books/NBK54100/

144. Griffen, R. M. (2024). *Bromelain.* Retrieved from Web MD: https://www.webmd.com/diet/supplement-guide-bromelain-bromelin

145. (2008). *The Liver.* Retrieved from Gastrointestinal Society & Canadian Society of Intestinal Research: https://badgut.org/information-centre/a-z-digestive-topics/the-liver-an-amazing-organ/

146. The Medical Biochemistry Page. (2025). *Bile Acid Synthesis, Metabolism, and Biological Functions.* Retrieved from The Medical Biochemistry Page: https://themedicalbiochemistrypage.org/bile-acid-synthesis-metabolism-and-biological-functions/

147. Sundaram, S. S. (2008). The multiple facets of ABCB4 (MDR3) deficiency. *Current Treatment Options in Gastroenterology.* doi:https://doi.org/10.1007/s11938-007-0049-4

148. Soret, P. L. (2023). Obeticholic acid as a second-line treatment for low phospholipid-associated cholelithiasis syndrome. *Alimentary Pharmacology & Therapeutics, 59*(1). doi: https://doi.org/10.1111/apt.17761

149. Sepe V, F. C. (2016). Insights on FXR selective modulation. Speculation on bile acid chemical space in the discovery of potent and selective agonists. *Scientific Reports, 6.* Retrieved from https://www.nature.com/articles/srep19008

150. Petrescu, A. K. (2018). Hypothalamus-Pituitary-Adrenal Dysfunction in Cholestatic Liver Disease. *Frontiers in Endocrinology, 9.* doi:https://doi.org/10.3389/fendo.2018.00660

151. Fenech, M. (2020). Chapter 4 - The Role of Nutrition in DNA Replication, DNA Damage Prevention and DNA Repair. *Principles of Nutrigenetics and Nutrigenomics.* doi:https://doi.org/10.1016/B978-0-12-804572-5.00004-5

152. D'Esposito, V. D. (2022). Lifestyle and Dietary Habits Affect Plasma Levels of Specific Cytokines in Healthy Subjects. *Frontiers in Nutrition, 9.* doi:https://doi.org/10.3389/fnut.2022.913176

153. Campbell WW, N. D. (2023). Nutritional Interventions: Dietary Protein Needs and Influences on Skeletal Muscle of Older Adults. *The Journals of Gerontology, Series A, 78*(1). doi:https://doi.org/10.1093/gerona/glad038

154. Furst, P. Y. (2000). *Conditionally Indispensable Amino Acids (Glutamine, Cyst(e)ine, Tyrosine, Arginine, Ornithine, Taurine) in Enteral Feeding and the Dipeptide Concept.* Retrieved from 154. Nestle Nutrition: https://www.nestlenutrition-institute.org/sites/default/files/documents-library/publications/secured/3c278e71ddaeb52300375aa29e389a4f.pdf

155. Deters, B. S. (2021). The role of glutamine in supporting gut health and neuropsychiatric factors. *Food Science and Human Wellness, 10*(2). doi:https://doi.org/10.1016/j.fshw.2021.02.003

156. Jong CJ, S. P. (2021). The Role of Taurine in Mitochondria Health: More Than Just an Antioxidant. *Molecules, 26*(16). doi:https://doi.org/10.3390/molecules26164913

157. Walter, Z. (2023). *Taurine: A Breakthrough Anti-Aging Intervention?* Retrieved from A4M Medicine Redefined: https://blog.a4m.com/taurine-a-breakthrough-anti-aging-intervention/

158. Childs, W. D. (2025). *L-Tyrosine and How it Benefits Thyroid & Adrenal Function: Should you Use it?* Retrieved from Dr. Childs: https://www.restartmed.com/l-tyrosine/

159. Semeco, A. A. (2024). *9 Benefits and Side Effects of Coenzyme Q10 (COQ10).* Retrieved from Healthline: https://www.healthline.com/nutrition/coenzyme-q10

16. Brian St. Pierre, M. R. (n.d.). *Settling the great grain debate.* Retrieved from Precision Nutrition: https://www.precisionnutrition.com/grain-wheat-debate

160. Burri L, H. N. (2012). Marine Omega-3 Phospholipids: Metabolism and Biological Activities. *International Journal of Molecular Sciences, 13*(11). doi:https://doi.org/10.3390/ijms131115401

161. Christie, W. (2025). *Phosphatidylethanolamine and Related Lipids.* Retrieved from The Lipid Web: https://lipidmaps.org/resources/lipidweb/lipidweb_html/lipids/complex/pe/index.htm

162. Züge, L. M. (2017). Use of avocado phospholipids as emulsifier. *LWT - Food Science and Technology, 79.* doi:https://doi.org/10.1016/j.lwt.2017.01.013

163. Cui, L. E. (2015). Phospholipids in Foods: Prooxidants or Antioxidants? *Journal of the Science of Food and Agriculture, 96*(1). doi: https://doi.org/10.1002/jsfa.7320

164. Wikipedia. (n.d.). *Phosphatidylethanolamine N-methyltransferase.* Retrieved from Wikipedia: https://en.wikipedia.org/wiki/Phosphatidylethanolamine_N-methyltransferase

165. Urikura M, M. J. (2012). Phosphatidic Acid Production in the Processing of Cabbage Leaves. *Journal of Agricultural and Food Chemistry, 60*(45). doi:https://doi.org/10.1021/jf303515z

166. Doniparthi, K. D. (2024). *How mRNA Vaccine Cell Damage Can Be Reversed Through Lipid Replacement Therapy.* Retrieved from EpochTV: https://www.theepochtimes.com/epochtv/how-mrna-vaccine-cell-damage-can-be-reversed-through-lipid-replacement-therapy-5635786?utm_source=Enews&src_src=Enews&utm_campaign=etv1-2024-05-02&src_cmp=etv1-2024-05-02&utm_medium=email&est=AAAAAAAAAAAAAAAAa%2B0jeyw

167. Johnson, J. (2019). *How does taurine affect the body?* Retrieved from Medical News Today: https://www.medicalnewstoday.com/articles/326714#taurine-roles-in-the-body

168. Thirupathi A, P. R. (2020). Taurine Reverses Oxidative Damages and Restores the Muscle Function in Overuse of Exercised Muscle. *Frontiers in Physiology, 11.* doi:https://doi.org/10.3389/fphys.2020.582449

169. Lee CW, L. T. (2023). The Effect of Choline and Resistance Training on Strength and Lean Mass in Older Adults. *Nutrients, 15*(18). doi:https://doi.org/10.3390/nu15183874

170. National Institutes of Health. (2022). *Choline.* Retrieved from National Institutes of Health - Office of Dietary Supplements: https://ods.od.nih.gov/factsheets/Choline-HealthProfessional/

171. Berg, D. (2018). Retrieved from Dr. Eric Berg DC: https://www.youtube.com/watch?v=rjVXFqiPDwE&t=289s

172. My Food Data. (n.d.). *Nutrient Ranking Tool - Thiamine (B1).* Retrieved from My Food Data: https://tools.myfooddata.com/nutrient-ranking-tool/thiamin-b1/beans-and-lentils+dairy-and-egg-products+fish+fruits+grains-and-pasta+meats+nuts-and-seeds+vegetables/highest/grams/common/no

173. Hill, A. (2023). *10 Types of Magnesium (and What to Use Each For).* Retrieved from Healthline: https://www.healthline.com/nutrition/magnesium-types#10.-Magnesium-orotate https://conceivehealth.com/blog/magnesium-fertility-everything-you-need-to-know/

174. Overton, E. (2022). *Why Thiamine Supplementation Requires Magnesium.* Retrieved from Magnesium: https://www.hormonesmatter.com/why-thiamine-supplementation-requires-magnesium/

175. Berg, D. E. (2025). *What Happens if You Stop Eating Sugar for 14 Days.* Retrieved from Dr. Berg: https://www.drberg.com/blog/why-do-nutrients-get-depleted-when-eating-sugar

176. Xiang, S. L. (2023). Increased Dietary Niacin Intake Improves Muscle Strength, Quality, and Glucose Homeostasis in Adults over 40 Years of Age. *The Journal of nutrition, health and aging, 27*(9). doi:https://doi.org/10.1007/s12603-023-1967-0

177. Vinmec. (2025). *9 Benefits of Niacin (Vitamin B3).* Retrieved from Vinmec: https://www.vinmec.com/eng/blog/9-science-based-benefits-of-niacin-vitamin-b3-en

178. Geisinger. (2023). *Nature's superhero: The benefits of vitamin C.* Retrieved from Geisingert: https://www.geisinger.org/health-and-wellness/wellness-articles/2022/05/27/19/57/natures-superhero-the-benefits-of-vitamin-c

179. Arora, D. S. (2025). *The best adrenal support supplements: How to heal your adrenal fatigue.* Retrieved from Womens Health Network: https://www.womenshealthnetwork.com/adrenal-fatigue-and-stress/supplements-for-adrenal-fatigue/

180. Brown, K. M. (2024). *5 Surprising Benefits of Zinc Sexually.* Retrieved from Hims: https://www.ncbi.nlm.nih.gov/pmc/articles/PMC9139143/

181. Gibney, J. W. (2003). Testosterone enhances the effect of growth hormone (GH) to increase IGF-I but exerts an anabolic effect that is independent of GH action. *Endocrine Abstracts, 5.* Retrieved from Endocrine Abstracts: https://www.endocrine-abstracts.org/ea/0005/ea0005p161

182. Kubala, J. (2022). *Zinc: Everything You Need to Know.* Retrieved from Healthline: https://www.healthline.com/nutrition/zinc

183. Rozenberg JM, K. M. (2022). The Role of the Metabolism of Zinc and Manganese Ions in Human Cancerogenesis. *Biomedicines, 10*(5). doi:https://doi.org/10.3390/biomedicines10051072

184. Yildiz A, K. Y. (2019). Effect of the Interaction Between Selenium and Zinc on DNA Repair in Association With Cancer Prevention. *Journal of Cancer Prevention, 24*(3). doi:https://doi.org/10.15430/JCP.2019.24.3.146

185. Stong, C. (2022). *ACE+TUDCA Therapy Effective for Reducing Clinical Symptoms of Gallstones.* Retrieved from Gastroenterology Advisor: https://www.gastroenterologyadvisor.com/gallbladder/acetudca-therapy-effective-for-reducing-clinical-symptoms-of-gallstones/

186. Song G, W. F. (2023). Potential therapeutic action of tauroursodeoxycholic acid against cholestatic liver injury via hepatic Fxr/Nrf2 and CHOP-DR5-caspase-8 pathway. *Clinical Science, 137*(7). doi:https://doi.org/10.1042/CS20220674

187. Vang S, L. K. (2014). The Unexpected Uses of Urso- and Tauroursodeoxycholic Acid in the Treatment of Non-liver Diseases. *Global Advances in Integrative Medicine and Health.* doi:https://doi.org/10.7453/gahmj.2014.017

188. Gunnars, C. (2023). *Berberine – A Powerful Supplement With Many Benefits.* Retrieved from Healthline: https://www.healthline.com/nutrition/berberine-powerful-supplement#how-it-works

189. Herzig S, S. R. (2017). AMPK: guardian of metabolism and mitochondrial homeostasis. *Nature Reviews Molecular Cell Biology , 19.* doi:https://doi.org/10.1038/nrm.2017.95

190. Zhu Y, X. N. (2022). Apoptosis Induction, a Sharp Edge of Berberine to Exert Anti-Cancer Effects, Focus on Breast, Lung, and Liver Cancer. *Frontiers in Pharmacology, 13.* doi:https://doi.org/10.3389/fphar.2022.803717

191. Robbins, J. (n.d.). *Abkhazia: Ancients of the Caucasus (Healthy at 100).* Retrieved from Abkhaz World: https://abkhazworld.com/aw/abkhazians/who-are-they/659-abkhazia-ancients-of-the-caucasus

192. Fuentes JM, M. P. (2024). The Role of Cardiolipin in Mitochondrial Function and Neurodegenerative Diseases. *Cells, 13*(7). doi:https://doi.org/10.3390/cells13070609

193. Creationwiki. (2025). *Sinites.* Retrieved from Creationwiki: https://creationwiki.org/Sinites

194. Lewis, R. (n.d.). *Does Chinese Civilization Come From Ancient Egypt?* Retrieved from Foreign Policy: . https://foreignpolicy.com/2016/09/02/did-chinese-civilization-come-from-ancient-egypt-archeological-debate-at-heart-of-china-national-identity/

195. New World Encyclopedia. (n.d.). *Three Sovereigns and Five Emperors.* Retrieved from New World Encyclopedia: https://www.newworldencyclopedia.org/entry/Three_Sovereigns_and_Five_Emperors

196. China Sage. (n.d.). *Ancient Chinese gods and rulers.* Retrieved from China Sage: https://www.chinasage.info/legendary-rulers.htm

197. Vasilev, D. (2021). *The Three Sovereigns - Fuxi, Shennong, and Huangdi.* Retrieved from My Myth Stories: https://mymythstories.com/three-sovereigns/

198. Truax, E. A. (1991). *Genesis According to the Miao People.* Retrieved from Institute for Creation Research: . https://www.icr.org/article/genesis-according-miao-people/

199 Asia Harvest. (n.d.). *The Mysterious Origins of the Miao.* Retrieved from Asia Harvest: https://www.asiaharvest.org/china-resources/guizhou/the-mysterious-origins-of-the-miao

200. Britannica. (2025). *Herbal Therapy - History of Chinese medicine.* Retrieved from Britannica: https://www.britannica.com/science/traditional-Chinese-medicine/Fu-Xi-and-the-bagua

201. Fagan. (2024). *God's Grace and Favor: The Spiritual Meaning of Number 5 in the Bible.* Retrieved from Increasing Faith Intl.: https://increasingfaithintl.org/biblical-meaning-of-number-5/

202. TCM World Foundation. (n.d.). *About Qi - Everything Is Energy.* Retrieved from Traditional Chinese Medicine World Foundation: https://www.tcmworld.org/what-is-tcm/about-qi/

203. TCM World Foundation. (n.d.). *Yin / Yang Theory - Encompassing Everything in the Universe.* Retrieved from Traditional Chinese Medicine World Foundation: https://www.tcmworld.org/what-is-tcm/yin-yang-theory/

204. The FengShui Academy. (n.d.). Retrieved from The FengShui Academy: https://www.thefengshuiacademy.com/5-elements/#1529998008821-f680a118-6071

205. Wikipedia. (n.d.). Retrieved from Wikipedia: https://en.wikipedia.org/wiki/Rain_dust

206. Schoenbart, B. S. (n.d.). *Yin Organs Overview, Yang Organs Overview.* Retrieved from How Stuff Works: https://health.howstuffworks.com/wellness/natural-medicine/chinese/yin-organs.htm?s1sid=hn0azuzspvrybtuvxchx5bnk&srch_tag=e6n6b23fz6urv4pkmchnu6hyjokcv4pp

207. Zhang L, Z. Q. (2011). The TCM Pattern of the Six-Zang and Six-Fu Organs Can Be Simplified into the Pattern of Five-Zang and One-Fu Organs. *Journal of Traditional Chinese Medicine, 31*(2). doi:https://doi.org/10.1016/S0254-6272(11)60030-7

208. Medline Plus. (2024). Hepatic venous circulation. *Medline Plus* . Retrieved from https://medlineplus.gov/ency/imagepages/19956.htm

209. TCM World Foundation. (n.d.). *Five Major Organs.* Retrieved from Traditional Chinese Medicine World Foundation: https://www.tcmworld.org/what-is-tcm/the-five-major-organ-systems/

210. . (n.d.). *Biochemical Oxygen Demand.* Retrieved from Jrank: https://science.jrank.org/pages/856/Biochemical-Oxygen-Demand.html

211. (n.d.). *Chapter 5: REGIONAL CIRCULATIONS.* Retrieved from Columbia.edu: http://www.columbia.edu/~kj3/Chapter5.htm

212. Felicia. (n.d.). *What are the 5 Element Theory Emotions? The Master Key to Understanding Yourself.* Retrieved from Guided by Elements: https://guidedbyelements.com/about-the-five-elements/symbolism-characteristics-and-associations/what-are-the-5-element-theory-emotions/

213. Bethany, J. P. (2024). *The Stages of Grief and What to Expect.* Retrieved from Healthline: https://www.healthline.com/health/stages-of-grief#5-stages

214. TCM World Foundation. (n.d.). *Meridians.* Retrieved from Traditional Chinese Medicine World Foundation: https://www.tcmworld.org/what-is-tcm/meridians/

215. Bhanote, M. M. (2023). *The Five Elements: What Science Has to Say About This Chinese Medicine Theory.* Retrieved from Healthline: https://www.healthline.com/health/mind-body/what-are-the-five-elements#in-tcm

216. Axe, D. (2023). *Top 15 Chinese Herbs & Superfoods to Boost Health.* Retrieved from Dr. Axe: https://draxe.com/nutrition/top-chinese-herbs-and-superfoods/

217. Adhikari, S. (2024). *Top 10 Traditional Ancient Chinese Foods.* Retrieved from AncientHistoryLists: https://www.ancienthistorylists.com/china-history/top-10-traditional-ancient-chinese-foods/

218. Kotyk, J. (2021). *Dairy: the deep historical link between China and Japan.* Retrieved from The China Story: https://www.thechinastory.org/dairy-the-deep-historical-link-between-china-and-japan/

219. Lawrence, F. (n.d.). *Can the world quench China's bottomless thirst for milk?* Retrieved from The Guardian: https://www.theguardian.com/environment/2019/mar/29/can-the-world-quench-chinas-bottomless-thirst-for-milk

220. BiblicalTraining.org,. (n.d.). *Testaments of the Twelve Patriarchs.* Retrieved from BiblicalTraining.org: https://www.biblicaltraining.org/library/testaments-of-the-twelve-patriarchs

221. Johnson, K. T. (2017). *Ancient Testaments of the Patriarchs.* Retrieved from Prophecy Watchers: https://prophecywatchers.com/product/ancient-testaments-patriarchs-new-ken-johnson/

222. Andrews, E. (n.d.). *What Do We Know About the Ancient Armenian Version of the Bible?* Retrieved from Christian Publishing House Blog: https://christianpublishinghouse.co/2017/04/19/the-armenian-version/

223. Rendsburg, G. (2021). *Ur Kasdim: Where Is Abraham's Birthplace?* Retrieved from The Torah - com: https://www.thetorah.com/article/ur-kasdim-where-is-abrahams-birthplace

224. The Firmament. (n.d.). *The Book of Jubilees.* Retrieved from The Firmament: https://www.thefirmament.org/scripture/book-of-jubilees/#h-chapter-10

225. Knight, K. (2023). *The Testaments of the Twelve Patriarchs.* Retrieved from New Advent: https://www.newadvent.org/fathers/0801.htm

226. The Firmament. (n.d.). *The Testaments of the Twelve Patriarchs of the House of Israel.* Retrieved from The Firmament: https://www.thefirmament.org/wp-content/uploads/pdf/Testaments%20of%20the%20Patriarchs.pdf

227. Sharma V, S. I. (2014). Acorus calamus (The Healing Plant): a review on its medicinal potential, micropropagation and conservation. *Natural Product Research, 28*(18). doi:https://doi.org/10.1080/14786419.2014.915827

228. Ma L, Z. J.-H. (2022). Effects of spleen nerve denervation on depression–like phenotype, systemic inflammation, and abnormal composition of gut microbiota in mice after administration of lipopolysaccharide: A role of brain–spleen axis. *Journal of Affective Disorders, 317.* doi:https://doi.org/10.1016/j.jad.2022.08.087

229. Caesar, S. (2006). *The Wealth and Power of the Biblical Patriarchs.* Retrieved from Associates for Biblical Research: https://biblearchaeology.org/research/patriarchal-era/2506-the-wealth-and-power-of-the-biblical-patriarchs

230. YAHWEH's Sword. (n.d.). *Food In The Time Of Abram.* Retrieved from YAHWEH's Sword: http://www.yahwehswordarchives.org/s-abraham/17_food_in_the_time_of_abram.htm

231. Venn-Watson S, S. N. (2023). Pentadecanoic Acid (C15:0), an Essential Fatty Acid, Shares Clinically Relevant Cell-Based Activities with Leading Longevity-Enhancing Compounds. *Nutrients, 15*(21). doi:https://doi.org/10.3390/nu15214607

232. Chin, K. R. (2024). *8 Raw Honey Benefits for Health.* Retrieved from Healthline: https://www.healthline.com/health/food-nutrition/top-raw-honey-benefits

233. Singh, A. N. (2024). *Women's Health Daily Health Is it healthy to mix milk with honey? Let's find out!* Retrieved from Health Shots: https://www.healthshots.com/daily-health/is-milk-with-honey-safe/

234. Nelte, F. (2016). *WHAT DOES THE BIBLE SAY ABOUT EATING FAT?* Retrieved from FrankNelte.com: https://www.franknelte.net/article.php?article_id=354

235. Patrick, D. R. (2025). *Neu5Gc.* Retrieved from Found My Fitness: https://www.foundmyfitness.com/topics/neu5gc

236. Failla S, C. M. (2023). Variability of sialic acids in meat from alternative species to beef and pork. *Animal Frontiers, 13*(6). doi:https://doi.org/10.1093/af/vfad058

237. (2014). Quantitative analysis of sialic acids in Chinese conventional foods by HPLC-FLD. *Open Journal of Preventative Medicine, 4*(2). doi:10.4236/ojpm.2014.42009

238. Barnett JA, G. D. (2020). Separating the Empirical Wheat From the Pseudoscientific Chaff: A Critical Review of the Literature Surrounding Glyphosate, Dysbiosis and Wheat-Sensitivity. *Frontiers in Microbiology, 11.* doi: https://doi.org/10.3389/fmicb.2020.556729

239. Oestermann, M. (2021). *Light & Olive Oil in the Time of the Bible.* Retrieved from Jerusalem Artists' Company: https://jerusalemartists.com/light-olive-oil-in-the-time-of-the-bible/

240. Fernando, I. (2023). *What Is Extra Virgin Olive Oil, and Why Is It Healthy?* Retrieved from Healthline: https://www.healthline.com/nutrition/extra-virgin-olive-oil#other-benefits

241. Quarles, C. (2021). *Was New Testament Wine Alcoholic?* Retrieved from The Center for Faith and Culture: https://cfc.sebts.edu/faith-and-culture/was-new-testament-wine-alcoholic/

242. Meng X, Z. J. (2020). Health Benefits and Molecular Mechanisms of Resveratrol: A Narrative Review. *Foods, 9*(3). doi:https://doi.org/10.3390/foods9030340

243. Grivennikov SI, G. F. (2010). Immunity, Inflammation, and Cancer. *Cell, 140*(6). doi:doi: 10.1016/j.cell.2010.01.025

244. Christovich A, L. X. (2022). Gut Microbiota, Leaky Gut, and Autoimmune Diseases. *Frontiers in Immunology, 13.* doi: https://doi.org/10.3389/fimmu.2022.946248

245. Guzior DV, Q. R. (2021). Review: microbial transformations of human bile acids. *Microbiome*(140). doi:10.1186/s40168-021-01101-1

246. Dmytriv TR, S. K. (2024). Intestinal barrier permeability: the influence of gut microbiota, nutrition, and exercise. *Frontiers in Physiology, 15.* doi:https://doi.org/10.3389/fphys.2024.1380713

247. Wolska A, D. R. (2018). Apolipoprotein C-II: New findings related to genetics, biochemistry, and role in triglyceride metabolism. *Atherosclerosis, 267.* doi:10.1016/j.atherosclerosis.2017.10.025

248. Froyen, E. (2021). The effects of fat consumption on low-density lipoprotein particle size in healthy individuals: a narrative review. *Lipids in Health and Disease*(86). Retrieved from https://lipidworld.biomedcentral.com/articles/10.1186/s12944-021-01501-0

249. Hughes, J. M. (2022). *Foods That Activate Brown Fat for Healthier Metabolism.* Retrieved from The Nutrition Insider: https://thenutritioninsider.com/fitness-weight-loss/foods-that-activate-brown-fat-for-healthier-metabolism/

250. Moss JW, R. D. (2016). Nutraceutical therapies for atherosclerosis. *Nature Reviews Cardiology, 13*(9). doi: 10.1038/nrcardio.2016.103

251. McCaddon A, M. J. (2023). Homocysteine—a retrospective and prospective appraisal. *Frontiers in Nutrition, 10.* doi:https://doi.org/10.3389/fnut.2023.1179807

252. Pfalzer AC, C. S.-F. (2014). S-adenosylmethionine mediates inhibition of inflammatory response and changes in DNA methylation in human

macrophages. *Physiological Genomics, 46*(17). doi:https://doi.org/10.1152/physiolgenomics.00056.2014

253. Liu Y, D. M. (2020). Trimethylamine N-Oxide Generated by the Gut Microbiota Is Associated with Vascular Inflammation: New Insights into Atherosclerosis. *Mediators of Inflammation.* doi: https://doi.org/10.1155/2020/4634172

254. Beckers, Bread. (2022). *Only Real Bread - Staff of Life.* Retrieved from Bread Beckers: https://www.youtube.com/watch?v=43s0MWGrlT8&t=32s

255. Becker, S. (2017). *Phytic Acid - Friend or Foe?* Retrieved from BreadBeckers: https://support.breadbeckers.com/kb/article/26-Phytic-Acid-Friend-or-Foe

256. Frey, M. (2021). *Kamut Nutrition Facts and Health Benefits.* Retrieved from Very Well: https://www.verywellfit.com/kamut-nutrition-facts-and-health-benefits-5093617

257. Devjc, S. (2022). *Grass-Fed Milk: Everything You Need to Know.* Retrieved from Healthline: https://www.healthline.com/nutrition/grass-fed-milk#environment-ethics

258. Gao Y, L. Y. (2025). Fermented Dairy Products as Precision Modulators of Gut Microbiota and Host Health: Mechanistic Insights, Clinical Evidence, and Future Directions. *Foods, 14*(11). doi:https://doi.org/10.3390/foods14111946

259. Mohol P, G. A. (2025). Blue Zone Dietary Patterns, Telomere Length Maintenance, and Longevity: A Critical Review. *Current Research in Nutrition and Food Science.* doi:https://dx.doi.org/10.12944/CRNFSJ.13.2.6

260. Uribarri J, W. S. (2010). Advanced Glycation End Products in Foods and a Practical Guide to Their Reduction in the Diet. *Journal of the Academy of Nutrition and Dietetics.* doi:10.1016/j.jada.2010.03.018

261. Ni Y, L. M. (2022). The Role of Gut Microbiota-Bile Acids Axis in the Progression of Non-alcoholic Fatty Liver Disease. *Trends in Microbiology, 31*(3). doi:https://doi.org/10.3389/fmicb.2022.908011

262. Anand S, M. S. (2022). Host-microbiome interactions: Gut-Liver axis and its connection with other organs. *Biofilms and Microbiomes, 8*(89). Retrieved from https://www.nature.com/articles/s41522-022-00352-6

263. Reuben, A. (2024). The Liver in Judaism. *Clinical Liver Disease, 23*(1). doi:10.1097/CLD.0000000000000234

264. Hajj-Boutros G, K. A. (2022). Potential mechanisms involved in regulating muscle protein turnover after acute exercise: A brief review. *Frontiers in Physiology, 13.* doi:https://doi.org/10.3389/fphys.2022.1106425

265. Hu C, X. Y. (2024). Effect of resistance circuit training on comprehensive health indicators in older adults: a systematic review and meta-analysis. *Scientific Reports*(8823). Retrieved from https://www.nature.com/articles/s41598-024-59386-9

266. Nuñez TP, A. F. (2020). Metabolic effects of two high-intensity circuit training protocols: Does sequence matter? *Journal of Exercise Science & Fitness, 18*(1).

267. Green, D. J. (2018). *ACE-SPONSORED RESEARCH: Is HIIT Resistance Exercise Superior to Traditional Resistance Training?* Retrieved from American Council on Exercise: https://www.acefitness.org/continuing-education/certified/june-2018/7011/ace-sponsored-research-is-hiit-resistance-exercise-superior-to-traditional-resistance-training/

268. Anand S, M. S. (2022). Host-microbiome interactions: Gut-Liver axis and its connection with other organs. *npj biofilms and microbiomes*(89). Retrieved from https://www.nature.com/articles/s41522-022-00352-6

269. Joung JY, C. J. (2019). A literature review for the mechanisms of stress-induced liver injury. *Brain and Behaviour, 9*(3). doi: https://doi.org/10.1002/brb3.1235

270. DiagnosTechs. (2016). *Phosphatidylserine: Cell Membrane Nutrient for Stress Resiliency.* Retrieved from DiagnosTechs: https://www.diagnostechs.com/2016/08/05/phosphatidylserine/

271. Patel AK, R. V. (2025). Physiology, Sleep Stages. *The National Center for Biotechnology.* Retrieved from https://www.ncbi.nlm.nih.gov/books/NBK526132/

272. Heli Z, H. C. (2022). Recent advances of γ-aminobutyric acid: Physiological and immunity function, enrichment, and metabolic pathway. *Frontiers in Nutrition, 9.* doi:https://doi.org/10.3389/fnut.2022.1076223

273. Xie X, D. D. (2022). Melatonin biosynthesis pathways in nature and its production in engineered microorganisms. *Synthetic and Systems Biotechnology, 7*(1). doi:https://doi.org/10.1016/j.synbio.2021.12.011

274. Group, E. (2017). *Top 5 Foods for the Pineal Gland.* Retrieved from Global Healing: https://globalhealing.com/blogs/education/foods-for-the-pineal-gland

275. Suni, E. (2025). *Foods That Help You Sleep.* Retrieved from Sleep Foundation: https://www.sleepfoundation.org/nutrition/food-and-drink-promote-good-nights-sleep

276. Coleman, D. S. (2023). *Chiropractic Care For Bloating and Digestion.* Retrieved from ChiroSolutions Center: https://mychirosolutions.com/chiropractic-care-bloating-and-digestion/

277. Kovanur Sampath K, M. R. (2015). Measureable changes in the neuro-endocrinal mechanism following spinal manipulation. *Medical Hypotheses, 85*(6). doi:https://doi.org/10.1016/j.mehy.2015.10.003

278. Sim, G. (n.d.). *Cold Plunge and Sauna: A Synergistic Longevity Therapy.* Retrieved from Mito.

279. González-Muñoz A, C.-C. M.-M.-N.-H.-G.-L. (2023). Efficacy of Photobiomodulation Therapy in the Treatment of Pain and Inflammation: A Literature Review. *Healthcare, 11*(7). doi:https://doi.org/10.3390/healthcare11070938

280. Park MK, L. W. (2018). Fat Intake and Risk of Skin Cancer in U.S. Adults. *Cancer Epidemiology, Biomarkers & Prevention, 27*(7). doi:https://doi.org/10.1158/1055-9965.EPI-17-0782

281. Koniver, L. (2023). Practical applications of grounding to support health. *Biomedical Journal, 46*(1).

282. Giulivi C, K. R. (2025). Earthing effects on mitochondrial function: ATP production and ROS generation. *FEBS OpenBio.* doi:https://doi.org/10.1002/2211-5463.70062

283. Tsui, H. S. (2019). *The roles of the Coq10 chaperone protein, cardiolipin, and endoplasmic reticulum-mitochondria contact sites in coenzyme Q biosynthesis and function.* Retrieved from eScholarship UCLA: https://escholarship.org/uc/item/5qw5b7hd

284. Palikaras K, T. N. (2014). Mitochondrial homeostasis: The interplay between mitophagy and mitochondrial biogenesis. *Experimental Gerontology, 56.* doi:https://doi.org/10.1016/j.exger.2014.01.021

285. Tesser JRP, C. A.-P. (2025). Vagus nerve-mediated neuroimmune modulation for rheumatoid arthritis: a pivotal randomized controlled trial. *Nature Medicine, 32.* doi:https://doi.org/10.1038/s41591-025-04114-7

286. Anar Dushpanova, e. a. (2025). Reconnecting the vagus nerve to the heart through nerve conduit preserves cardiac function in a minipig model of right cardiac vagotomy. *Science Translational Medicine, 17*(830). Retrieved from https://www.science.org/doi/10.1126/scitranslmed.aea4306

287. Kelly C, T. C. (2024). A platform to map the mind–mitochondria connection and the hallmarks of psychobiology: the MiSBIE study.

Trends in Endocrinology & Metabolism, 35(10). doi:https://doi.org/10.1016/j.tem.2024.08.006

288. Fagundes CP, W.-c. E. (2025). Psychological Science at the Cellular Level: Mitochondria's Role in Health and Behavior. *Current Directions in Psychological Science, 35*(2). doi:https://doi.org/10.1177/09637214251380214

289. Scheiblich H, E. F.-G. (2024). Microglia rescue neurons from aggregate-induced neuronal dysfunction and death through tunneling nanotubes. *Neuron, 112*(18). Retrieved from https://www.cell.com/neuron/fulltext/S0896-6273(24)00491-4

www.ingramcontent.com/pod-product-compliance
Lightning Source LLC
LaVergne TN
LVHW100254210926
848480LV00003B/920